The Country Life Movement in the United States

by Liberty Hyde Bailey

with an introduction by Roger Chambers

This work contains material that was originally published in 1920.

This publication was created and published for the public benefit, utilizing public funding and is within the Public Domain.

This edition is reprinted for educational purposes and in accordance with all applicable Federal Laws.

Introduction Copyright 2017 by Roger Chambers

Self Reliance Books

Get more historic titles on animal and stock breeding, gardening and old fashioned skills by visiting us at:

http://selfreliancebooks.blogspot.com/

Introduction

I am pleased to present yet another title on Gardening.

The work is in the Public Domain and is re-printed here in accordance with Federal Laws.

As with all reprinted books of this age that are intended to perfectly reproduce the original edition, considerable pains and effort had to be undertaken to correct fading and sometimes outright damage to existing proofs of this title. At times, this task is quite monumental, requiring an almost total "rebuilding" of some pages from digital proofs of multiple copies. Despite this, imperfections still sometimes exist in the final proof and may detract from the visual appearance of the text.

I hope you enjoy reading this book as much as I enjoyed making it available to readers again.

Roger Chambers

TO

Charles W. Garfield

—SEER OF VISIONS, PROPHET OF THE

BETTER COUNTRY LIFE—

I dedicate this budget

of opinions

CONTENTS

THE COUNTRY-LIFE MOVEMENT
Pages 1–3

It is not a back-to-the-land movement, 1 — This book, 2.

THE NATIONAL MOVEMENT
Pages 4–13

A transition period, 6 — The Commission on Country Life, 7 — The three fundamental recommendations of the Commission, 9 — A national conference of country life, 12 — A voluntary movement, 12 — The international phase, 13.

SOME INTERRELATIONS OF CITY AND COUNTRY
Pages 14–30

Some contrasts of town folk and country folk, 14 — Comparisons of town and country affairs, 16 — The two minds, 17 — Will the American farmer hold his own? 19 — The first two remedies, 21 — Movement from city to country as remedy, 23 — Sending the surplus population to the country, 25 — Back-to-the-village, 26 — Can a city man make a living on a farm? 27 — What the city may do, 30.

Contents

THE DECLINE IN RURAL POPULATION. — ABANDONED FARMS

Pages 31–43

Significance of the decline, 32 — The abandoned farms, 37 — The new farming, 41.

RECLAMATION IN RELATION TO COUNTRY LIFE; AND THE RESERVE LANDS

Pages 44–54

The interests of society in the work, 45 — A broad reclamation movement, 50 — Supplemental irrigation, 51 — We need reserves, 53.

WHAT IS TO BE THE OUTCOME OF OUR INDUSTRIAL CIVILIZATION?

Pages 55–60

(1) The making of a new society, 56 — (2) The fighting edge, 57.

THE FUNDAMENTAL QUESTION IN AMERICAN COUNTRY LIFE

Pages 61–84

Agriculture in the public schools, 62 — The American contribution, 65 — The dangers in the situation, 66 — The

Contents

present educational institutions, 68 — The need of plans to coördinate this educational work, 71 — Outline of a state plan, 72 — A state extension program, 75 — Special local schools for agriculture, 76 — The lessons of experience, 79.

WOMAN'S CONTRIBUTION TO THE COUNTRY-LIFE MOVEMENT

PAGES 85–96

The affairs of the household, 88 — The affairs of the community, 90 — The woman's outlook, 92 — The means of education, 93.

HOW SHALL WE SECURE COMMUNITY LIFE IN THE OPEN COUNTRY?

PAGES 97–133

Hamlet life, 100 — The category of agencies, 104 (increase of population, 105; dividing up of large farms, 106; assembling farms, 106; recreative life, 107; local politics, 108; rural government, 108; community program for health, 112; local factories and industries, 116; the country store, 118; the business men's organizations, 119; great corporations, 120; local institutions, 122; local rural press, 123; many kinds of extension teaching, 123; all kinds of communication, 124; economic or business coöperation, 125; personal gumption and guidance, 132) — Community interest is of the spirit, 133.

Contents

A POINT OF VIEW ON THE LABOR PROBLEM

Pages 134–148

Reasons for the labor question, 135 — The remedies, 137 — Public or social bearings, 139 — Supervision in farm labor, 142 — What is the farmer to do? 146.

THE MIDDLEMAN QUESTION

Pages 149–164

Farmer does not get his share, 149 — Relation of the question to cost-of-living, 153 — The farmer's part, 156 — The middleman's part, 157 — A system of economic waste, 158 — Coöperation of farmers will not solve it, 158 — It is the business of government, 160 — Must be a continuing process of control, 161.

COUNTY AND LOCAL FAIRS

Pages 165–177

Nature of the fair, 165 — Features to be eliminated, 167 — Constructive program, 167 — The financial support, 168 — An educational basis, 169 — Ask every person to prove up, 171 — Sports, contests, and pageants, 173 — Premiums, 174 — It is time to begin, 175 — The fair ground, 176 — My plea, 177.

Contents

THE COUNTRY-LIFE PHASE OF CONSERVATION

Pages 178–200

These subjects have a history, 180 — They are not party-politics subjects, 182 — The soil is the greatest of all resources, 183 — The soil crust, 185 — No man has a right to plunder the soil, 188 — Ownership *vs.* conservation, 190 — The philosophy of saving, 192 — The conservation of food, 194 — The best husbandry is not in the new regions, 196 — Another philosophy of agriculture, 197 — The obligation of the farmer, 198 — The obligation of the conservation movement, 200.

PERSONAL SUGGESTIONS

Pages 201–220

The open country must solve its own problems, 201 — Profitable farming is not a sufficient object in life, 202 — New country professions, 203 — The personal resources, 204 — The meaning of the environment, 205 — Historic monuments, 208 — Improvement societies, 209 — Entertainment, 211 (Music spirit, 212; drama, 213) — The business of farming, 217.

THE COUNTRY-LIFE MOVEMENT

The country-life movement is the working out of the desire to make rural civilization as effective and satisfying as other civilization.

It is not an organized movement proceeding from one center or even expressing one set of ideas. It is a world-motive to even up society as between country and city; for it is generally understood that country life has not reached as high development within its sphere as city life has reached within its sphere.

We call it a new subject. As a "movement," or a recognized set of problems needing attention, it may possibly be called new; but in reality it is new only to those who have recently discovered it.

It is not a back-to-the-land movement.

The country-life movement must be sharply distinguished from the present popular back-to-the-land agitation. The latter is primarily a city or town impulse, expressing the desire of townspeople to escape, or of cities to find relief, or of real estate dealers to sell land; and in part it is the result of the doubtful propaganda to decrease the cost of living by sending more persons to the land, on the mostly mistaken assumption that more products will thereby be secured for the world's markets.

The back-to-the-land agitation is not necessarily to be discouraged, yet we are not to expect more of it than it can accomplish; but whatever the outward movement to the land may be, the effort to effectualize rural society, for the people who now comprise this society, is one of the fundamental problems now before the people.

The country-life and back-to-the-land movements are not only little related, but in many ways they are distinctly antagonistic.

This book.

The foregoing paragraphs indicate the subject of this book. I mean only to express opinions on a few of the questions that are popularly under discussion, or that are specially important at this time. I shall present no studies, and I intend to follow no systematic course. Some of these subjects I have already discussed with the public, but they may now have new expression or relations.

The lack of adjustment between city and country must be remedied, but the remedies lie in fundamental processes and not in the treatment of symptoms. Undoubtedly very much can be done to even up the economic situation and the distribution of population; and this needs careful and continuous study by commissions or other agencies created for the purpose. We are scarcely in sight of the good that such agencies could accomplish. I hope that this book may suggest some of the things to be considered. The past century belonged to the city; the present century should belong also to agriculture and the open country.

THE NATIONAL MOVEMENT

THE present revival of rural interest is immediately an effort to improve farming; but at bottom it is a desire to stimulate new activity in a more or less stationary phase of civilization. We may over-exploit the movement, but it is sound at the center. For the next twenty-five years we may expect it to have great influence on the course of events, for it will require this length of time to balance up society. Politicians will use it as a means of riding into power. Demagogues and fakirs will take advantage of it for personal gain. Tradesmen will make much of it. Writers are even now beginning to sensationalize it.

But there will also arise countrymen with statesmanship in them; if not so, then we cannot make the progress that we need. The movement will have its significant political aspect, and we may look for governors of states and perhaps more than one President of the United

States to come out of it. In the end, the farmer controls the politics because he makes the crops on which the wealth of the country depends. There is probably a greater proportion of tax-payers among voting farmers than among city people.

Considered in total results, educational and political as well as social and economic, the country-life movement in North America is probably farther advanced than in any other part of the world. It may not have such striking manifestations in some special lines, and our people may not need so much as other peoples that these particular lines be first or most strongly attacked. The movement really has been under way for many years, but it has only recently found separate expression. Most of the progress has been fundamental, and will not need to be done over again. The movement is well afoot among the country people themselves, and they are doing some of the clearest thinking on the situation. Many of our own people do not know how far we have already come.

A transition period.

Such undercurrent movements are usually associated with transition epochs. In parts of the Old World the nexus in the social structure has been the landlord, and the change in land-tenure systems has made a social reorganization necessary. There is no political land-tenure problem in the United States, and therefore there is no need, on that score, of the coöperation of small owners or would-be owners to form a new social crystallization. But there is a land problem with us, nevertheless, and this is at the bottom of our present movement: it is the immanent problem of remaining more or less stationary on our present lands, rather than moving on to untouched lands, when the ready-to-use fertility is reduced. We have had a new-land society, with all the marks of expansion and shift. We are now coming to a new era; but, unlike new eras in some other countries, it is not complicated by hereditary social stratification. Our real agricultural development will now begin.

In the discussion of these rural interests, old foundations and old ideas in all probability will be torn up. We shall probably discard many of the notions that now are new and that promise well. We may face trying situations, but something better will come out of it. It is now a time to be conservative and careful, and to let the movement mature.

The commission on country life.

The first organized expression of the country-life movement in the United States was the appointment of the Commission on Country Life by President Roosevelt in August, 1908. It was a Commission of exploration and suggestion. It could make no scientific studies of its own within the time at its command, but it could put the situation before the people. President Roosevelt saw the country-life problem and attacked it.

The Commission made its Report to the President early in 1909. It found the general level of country life in the United States to be good as compared with that of any previous time, but yet "that agriculture is not commercially as profitable as it is entitled to be for the labor and energy that the farmer expends and the risks that he assumes, and that the social conditions in the open country are far short of their possibilities."

A dozen large reasons for this state of affairs, a state that directly curtails the efficiency of the nation, are given in the Report; and it suggests many remedies that can be set in motion by Congress, states, communities, and individuals. The three "great movements of the utmost consequence that should be set under way at the earliest possible time, because they are fundamental to the whole problem of ultimate permanent reconstruction" are: taking inventory of country life by means of "an exhaustive study or survey of all the conditions that surround the business of farming and the people who live in the country, in order to take stock of our resources and to supply the farmer with local knowledge"; the organizing of a nationalized extension work; the inauguration of a general campaign of rural progress.

It is suggested that Congress provide "some means or agency for the guidance of public opinion toward the development of a real rural society that shall rest directly on the land."

The Report of the Commission on Country Life makes no discussion of the city-to-country movement.

The Report recognizes the fundamental importance of the agricultural experiment stations and of the great chain of land-grant colleges and of government departments and of other agencies; and the work that it proposes is intended to be supplementary to them.

The three fundamental recommendations of the Commission.

The taking stock of the exact condition and materials of country life is immensely important, for we cannot apply remedies before we make a diagnosis, and an accurate diagnosis must rest on a multitude of facts that we do not now possess. This is the scientific rather than the doctrinaire, politics, and oracular method of approaching the subject. It is of the first importance that we do not set out on this new work with only general opinions and superficial and fragmentary knowledge. Every rural community needs to have a program of its own carefully worked out, and this program should rest on a physical valuation. It may be some time yet before the importance and magnitude of this undertaking will impress the minds of the people, but it is essential to the best permanent progress.

Agricultural extension work of a well-organized kind is now beginning to come out of the colleges of agriculture, and this must be extended and systematized so that, with other agencies, it may reach every last man on the land. A bill to set this work in motion is now before Congress.

The third recommendation of the Commission for immediate action is "the holding of local, state, and even national conferences on rural progress, designed to unite the interests of education, organization, and religion into one forward movement for the re-building of country life. Rural teachers, librarians, clergymen, editors, physicians, and others may well unite with farmers in studying and discussing the rural question in all its aspects. We must in some way unite all institutions, all organizations, all individuals having any interest in country life into one great campaign for rural progress."

Conferences are now being held in many parts of the Union by universities, colleges, state departments of agriculture, chambers of commerce, business organizations, and other bodies. This will make public opinion. Such conventions, discussing the larger social, political, and economic relations of country life, should now be held in every state and geographical region.

It is now time that states undertake country-life programs. There is still much attack of symptoms; but persons in political offices, for the most, are not yet well-enough informed to make the most of the rural situation as it exists, or to utilize to the best advantage the talent and the institutions that the country now possesses.

One has only to read the recommendations to legislative bodies to recognize the relative lack as yet of constructive plans for the improvement of rural conditions.

A national conference on country life.

If there should be state and local conferences for country life, so also should there be a national conference, meeting yearly. Such a conference should not be an agricultural convention in the ordinary sense, nor is it necessary that it be held in commanding agricultural regions. It should deal with the larger affairs and relations in their applications to rural civilization.

A voluntary movement.

The interest in country life is gradually assuming shape as a voluntary movement outside of government, as it properly should do. It should be in the best sense a popular movement; for if it is not a really popular movement, it can have little vitality, and exert little effect on the mass of the people. As it gets under motion, certain things will crystallize out of it for government to do; and governments will do them.

As a pure matter of propagation, such a voluntary organized movement would have the greatest value; for, in these days, simple publicity often accomplishes more than legislative action.

The international phase.

If the interest in rural economics and sociology is world wide, then we should have international institutions to represent it. Several organizations now represent or include certain phases. We need such an institution not so much for propaganda as for research. A Country Life Institute has been proposed by Sir Horace Plunkett, who is so well known and admired by all students of rural situations through his far-seeing work in Ireland and his many fruitful suggestions for America. It would seem that here is an unusual opportunity for a great and productive foundation.

SOME INTER-RELATIONS OF CITY AND COUNTRY

EVERY one knows that city populations are increasing more rapidly than country populations. By some persons, this of itself is considered to be a cause for much alarm. But the relative size of the populations is not so disturbing as the economic and social relations existing between these two phases of civilization.

Some contrasts of town folk and country folk.

We know that farming is the primitive and underlying business of mankind. As human desires have arisen, other occupations have developed to satisfy the increasing needs and aspirations, the products of the earth have been assembled and changed by manufacture into a thousand forms, and these departures have resulted in more refined products, a more resourceful civilization, and a more sensitive people.

Inter-relations of City and Country

Complex developments have been taken out of and away from agriculture, and have left it with the simple and undifferentiated products and the elemental contact with nature. The farmer is largely a residuary force in society; this explains his conservatism.

If we have very highly developed persons in the city, we have very rugged persons in the country. If the sense of brotherhood is highly evolved in the city, individualism is strongly expressed in the country. If the world-movement appeals to men in the city, local attachments have great power with men in the country. If commercial consolidation and organization are characteristic of the city, the economic separateness of the man or family is highly marked in the country. The more marked progress of the city is due to its greater number of leaders and to its consolidated interests; country people are personally as progressive as city people of equal intellectual groups, but they have not been able to attract as much attention or perhaps to make as much headway.

Comparisons of town and country affairs.

Civilization oscillates between two poles. At the one extreme is the so-called laboring class, and at the other are the syndicated and corporate and monopolized interests. Both these elements or phases tend to go to extremes. Many efforts are being made to weld them into some sort of share-earning or commonness of interest, but without very great results. Between these two poles is the great agricultural class, which is the natural balance-force or the middle-wheel of society. These people are steady, conservative, abiding by the law, and are to a greater extent than we recognize a controlling element in our social structure.

The man on the farm has the opportunity to found a dynasty. City properties may come and go, rented houses may be removed, stocks and bonds may rise and fall, but the land still remains; and a man can remain on the land and subsist with it so long as he knows how to handle it properly. It is largely, therefore, a question of education as to how long any family can establish itself on a piece of land.

In the accelerating mobility of our civilization it is increasingly important that we have many anchoring places; and these anchoring places are the farms.

These two phases of society produce marked results in ways of doing business. The great centers invite combinations, and, because society has not kept pace with guiding and correcting measures, immense abuses have arisen and the few have tended to fatten on the many. There are two general modes of correcting, or at least of modifying, these abuses,— by doing what we can to make men personally honest and responsible, and by evening up society so that all men may have something like a natural opportunity.

The two minds.

There is a town mind and a country mind. I do not pretend to know what may be the psychological processes, but it is clear that the mode of approach to the problem of life is very different as between the real urbanite and the real ruralite. This factor is not sufficiently taken into account by city men who would remove to real farms and make a living there. It is the cause of most of the failure of well-intentioned social workers to accomplish much for country people.

All this is singularly reflected in our literature, and most of all, perhaps, in guide-books. These books — made to meet the demand — illustrate how completely the open country has been in eclipse. There is little rural country discoverable in these books, unless it is mere "sights" or "places," — nothing of the people, of the lands, of the products, of the markets, of the country dorfs, of the way of life; but there is surfeit of cathedrals, of history of cities, of seats of famous personages, of bridges and streets, of galleries and works of art. We begin to see evidences of travel out into the farming regions, part of it, no doubt, merely a desire for new experiences and diversion, and we shall now look for guide-books that recognize the background on which the cities rest.

But all this will call for a new intention in travel.

Will the American farmer hold his own?

What future lies before the American farmer? Will he hold something like a position of independence and individualism, or will he become submerged in the social order, and form only an underlying stratum? What ultimate hope is there for a farmer as a member of society?

It is strange that the producer of the raw material has thus far in the history of the world taken a subordinate place to the trader in this material and to the fabricator of it. The trader and fabricator live in centers that we call cities. One type of mind assembles; the other type remains more or less scattered. So there have arisen in human society two divergent streams, — the collective and coöperative, and the isolated and individualistic.

The fundamental weakness in our civilization is the fact that the city and the country represent antagonistic forces. Sympathetically, they have been and are opposed. The city lives on the country. It always tends to destroy its province.

The city sits like a parasite, running out its roots into the open country and draining it of its substance. The city takes everything to itself — materials, money, men — and gives back only what it does not want; it does not reconstruct or even maintain its contributory country. Many country places are already sucked dry.

The future state of the farmer, or real countryman, will depend directly on the kind of balance or relationship that exists between urban and rural forces; and in the end, the state of the city will rest on the same basis. Whatever the city does for the country, it does also for itself.

Mankind has not yet worked out this organic relation of town and country. City and country are gradually coming together fraternally, but this is due more to acquaintanceship than to any underlying coöperation between them as equal forces in society. Until such an organic relationship exists, civilization cannot be perfected or sustained, however high it may rise in its various parts.

The first two remedies.

Of course there are no two or even a dozen means that can bring about this fundamental adjustment, but the two most important means are at hand and can be immediately put into better operation.

The first necessity is to place broadly trained persons in the open country, for all progress depends on the ability and the outlook of men and women.

The second necessity is that city folk and country folk work together on all great public questions. Look over the directories of big undertakings, the memberships of commissions and councils, the committees that lay plans for great enterprises affecting all the people, and note how few are the names that really represent the ideas and affairs of the open country. Note also how many are the names that represent financial interests, as if such interests should have the right of way and should exert the largest influence in determining public policies. In all enterprises and movements in which social benefit is involved, the agricultural country should be as much represented as the city. There are men and women enough out in the open country who are qualified to serve on such commissions and directories; but even if there were not, it would now be our duty to raise them up by giving rural people a chance. Rural talent has not had adequate opportunity to express itself or to make its contribution to the welfare of the world.

I know it is said, in reply to these remarks, that many of the city persons on such organizations are country-born, but this does not change the point of my contention. Many country-born townsmen are widely out of knowledge of present rural conditions, even though their sympathies are still countryward. It is also said that many of them live in country villages, small cities, and in suburbs; but even so, their real relations may be with town rather than country, and they may have little of the farm-country mind; and the suburban mind is really a town mind.

Every broad public movement should have country people on its board of control. Both urban and rural forces must shape our civilization.

Movement from city to country as a remedy.

Some persons seem to think that the movement of city men out to the country offers a solution of country problems. It usually offers only a solution of a city problem,— how a city man may find the most enjoyment for his leisure hours and his vacations. Much of the rising interest in country life on the part of certain people is only a demand for a new form of entertainment. These people strike the high places in the country, but they may contribute little or nothing to real country welfare. This form of entertainment will lose its novelty, as the sea-shore loses it for the mountains and the horse loses it for the motor-car or aëroplane. The farming of some city men is demoralizing to real country interests. I do not look for much permanent good to come to rural society from the moving out of some of the types of city men or from the farming in which they ordinarily engage.

I am glad of all movements to place persons on the land who ought to go there, and to direct country-minded immigrants away from the cities; but we must not expect too much from the process, and we must distinguish between the benefit that may accrue to these persons themselves and the need of reconstruction in the open country. It is one thing for a family to move to the land in order to raise its own supplies and to secure the benefits of country life; it is quite another thing to suppose that an exodus from city to country will relieve the economic situation or make any difference in the general cost of living, even assuming that the town folk would make good farmers. And we must be very careful not to confuse suburbanism and gardening with country life.

To have any continuing effect on the course of rural development, a person or an agency must become a real part and parcel of the country life.

Sending the surplus population to the country.

It is also proposed to send to the country the poor-to-do and the dissatisfied and the unemployed. This is very doubtful policy. In the first place, the presumption is that a person who does not do well or is much dissatisfied in the town would not do well in the country. In the second place, the country does not need him. We may need more farm labor, as we need more of all kinds of labor, but in the long run this labor should be produced mostly in the country and kept there by a profitable and attractive rural life. The city should not be expected permanently to supply it. The labor that the city can supply with profit to country districts is the very labor that is good enough for the city to keep.

The relief of cities, if relief is to be secured, must lie in the evolution of the entire situation, and not merely in sending the surplus population into the country.

In my opinion, the present back-to-the-farm cry is for the most part unscientific and unsound, as a corrective of social ills. It rests largely on the assumption that one solution of city congestion is to send people away from itself to the open country, and on another assumption that "a little farm well tilled" will abundantly support a family. There is bound to be a strong reaction against much of the present agitation. We are to consider the welfare of the open country as well as that of the city itself. The open country needs more good farmers, whether they are country-bred or city-bred; but it cannot utilize or assimilate to any great extent the typical urban-minded man; and the farm is not a refuge.

Back-to-the-village.

It seems to me that what is really needed is a back-to-the-village movement. This should be more than a mere suburban movement. The suburban development enlarges the boundaries of the city. It is perfectly feasible, however, to establish manufacturing and other concentrated enterprises in villages in many parts of the country. Persons connected with these enterprises could own small pieces of land,

and by working these areas could add something to their means of support, and also satisfy their desire for a nature-connection. In many of the villages there are vacant houses and comparatively unoccupied land in sufficient number and amount to house and establish many enterprises; and there would be room for growth. If the rural village, freed from urban influences, could then become a real integrating part of the open country surrounding it, all parties ought to be better served than now, and the social condition of both cities and country ought to be improved. We have over-built our cities at the expense of the hamlets and the towns. I look for a great development of the village and small community in the next generation; but this involves a re-study of freight rates.

Can a city man make a living on a farm?

Yes, if he is industrious and knows how. Many city persons have made good on the land, but they are the exceptions, unless they began young.

There is the most curious confusion of ideas on this question. We say that farming requires the highest kind of knowledge and at the same time think that any man may go on a farm, no matter how unsuccessful he may have been elsewhere. Even if he has been successful as a middleman or manufacturer or merchant, it does not at all follow that he would be successful as a farmer. Farming cannot be done at long range or by proxy any more than banking, or storekeeping, or railroading, and especially not by one who does not know how; and he cannot learn it out of books and bulletins. If a man can run a large factory without first learning the business, or a theater or a department store, then he might be able also to run a farm, although the running of a farm of equal investment would probably be the more difficult undertaking.

I am glad to see earnest city men go into farming when they are qualified to do so, but I warn my friends that many good people who go out from cities to farms with golden hopes will be sadly disappointed. Farming is a good business and it is getting better, but it is a business for farmers; and on the farmers as a group must rest the immediate responsibility of improving rural conditions in general.

The younger the man when he begins to consider being a farmer, the greater will be his chances of success; here the student has the great advantage.

City people must be on their guard against attractive land schemes. Now and then it is possible to pay for the land and make a living out of it at the same time, but these cases are so few that the intending purchaser would better not make his calculations on them. Farming is no longer a poor man's business. It requires capital to equip and run a farm as well as to buy it, the same as in other business. It is a common fault of land schemes to magnify the income, and to minimize both the risks and the amount of needed capital. Plans that read well may be wholly unsound or even impossible when translated into plain business practice. The exploiting of exceptional results in reporter's English and with charming pictures is having a very dangerous effect on the public mind; and even some of these results may not stand business analysis.

What the city may do.

It is not incumbent on cities, corporations, colleges, or other institutions to demonstrate, by going into general practical farming, that the farming business may be made to pay: thousands of farmers are demonstrating this every day.

If the city ever saves the open country, it will be by working out a real economic and social coördination between city and country, not by the city going into farming.

We need to correct the abnormal urban domination in political power, in control of the agencies of trade, in discriminatory practices, and in artificial stimulation, and at the same time to protect the evolution of a new rural welfare. The agrarian situation in the world is not to be met alone by increasing the technical efficiency of farming.

THE DECLINE IN RURAL POPULATION — ABANDONED FARMS

The decline in rural population grows out of economic conditions. Men move to the centers, where they can make the best living for themselves and families. It is difficult, however, for the farmer to "pull up stakes" and move. He is tied to his land. The result is that many men who really could do better in the town than on their farms are still remaining on the land. These persons will continue to remove to towns and cities as they are gradually forced off their lands; or if they are not forced off, their children will go, and the farm will eventually change hands.

Social reasons also have their influence in the movement of rural populations to towns. The social resources in the country in recent years have been very meager, because the social attractions of the towns have drawn away from the activities of the open country, and also more or less because the population itself is decreasing and does not allow, thereby, for so close social cohesion.

It is not to be expected that the counter-movement from the towns and cities to the open country will yet balance in numbers the movement of population from the country to the city.

It is important that conditions be so improved for the open country that those who are born on the farms and who are farm-minded shall feel that opportunities are at least as good for them there as in the city, and thereby prevent the exodus to the city or to other business of persons who really ought to remain in the rural regions.

Significance of the decline.

It is commonly assumed that a decline in rural population in any region is itself evidence of a real decline in agriculture. This conclusion, however, does not at all follow. The shift in population as between town and country is an expression of very many causes. In some cases it may mean a lessening in economic efficiency in the region, and in some cases an actual increase in such efficiency.

It must be remembered that we have been passing from the rural to the urban phase of civilization. The census of 1900 showed approximately one-third of our people on farms or closely connected with farms, as against something like nine-tenths a hundred years previous. It is doubtful whether we have yet struck bottom, although the rural exodus may have gone too far in some regions; and we may not permanently strike bottom for some time to come.

We think of Washington, Jefferson, Monroe, and other early patriots as countrymen, and we are likely to deplore the fact that countrymen no longer represent us in high places. The fact is that "the fathers" represented all society, because society in their day was not clearly differentiated between city and country. They were at the same time countrymen and city men, but the city was the incidental or secondary interest. To-day, the conditions are reversed. The city has come to be the preponderating force, and the country is largely incidental and secondary so far as the shaping of policies is concerned; but this does not prove that a greater ratio of country population is needed. The number of persons now living in the open country is probably sufficient, if the persons were all properly effective. The real problem before the American people is how to make the country population most effective, not how to increase this population; the increase will be governed by the operation of economic law.

The sorting of our people has not yet reached its limit of approximate stability. Many persons who live on the land really are not farmers, but are the remainders of the rural phase of society.

A decline in rural population in any region may be expressive of the general adjustment as between country and city; it may mean the passing out of active cultivation of large areas of land that ought to be in forest or in extensive systems of agriculture; it may mean the moving out of well-to-do farmers to cheaper lands, as an expression of the land-hunger of the American; it may be due in some cases to the retiring of well-to-do persons from the farms to the town; and other causes are at work in particular localities. The rural population of Iowa is decreasing, but the agricultural production and land valuation are increasing.

The lessened production of live-stock, of which we have recently heard so much, is probably not due to any great extent, if at all, to decreasing rural population. It is in part due to the shift in farming following the passing of the western ranges, and in part to the lack of a free market, and in part to a changing adjustment in farming practices. This situation will take care of itself if the markets are not manipulated or controlled.

Many publicists are alarmed at the lessened production of farm products in comparison with imports, and fear that the balance of trade will be seriously turned against us, with a rise in the rate of exchange. It is not to be expected that we shall maintain our former rate of export of raw crops, nor is it desirable from the point of view of maintaining the fertility of our lands that we should do so; but the maintenance of production is now to depend on farming every acre better, in larger farms as well as in smaller farms, rather than on taking up new acres.

The ultimate importance of agriculture to civilization, in other words, lies not in the number of persons it supports, but in the fact that it must continue to provide supplies for the populations of the earth when mining and exploitation are done, when there are no new lands, and when we shall have taken away all the first flush of the earth's bounty. The character of the farm man, therefore, becomes of supreme importance, and all the institutions of society must lend themselves to this personal problem.

We shall never again be a rural people. We want the cities to grow; and as they grow they should learn how to manage themselves. How they shall meet their questions of population is not my problem; and I have no suggestions to make on that subject.

The abandoned farms.[1]

If persons move from any part of the country until there is a marked absolute falling off in population, it must follow that certain lands shall be left unused, or shall be combined with adjacent lands into larger business units. It is no anomaly that there are " abandoned farms " (they are seldom really abandoned, but more or less unused), and it is natural that they should be in the remoter, hillier, and poorer regions. So are shop buildings abandoned on back streets, and likewise factories on lonely streams.

Some farms in the remote or difficult regions are still well utilized, because a skillful man has met the situation; others may be very nearly or quite disused; between the two extremes there is every shade of condition. Some farms are falling into disuse for one reason and some for another. In some cases, it is because the family is merely broken up and is moving

[1] Another discussion of this subject may be found in " The State and the Farmer."

off. In other cases, it is because the farm can no longer make a good living for a man and his family, giving him the things that a man of the twentieth century wants. A farm in the hill region that was large enough to support a man fifty or seventy-five years ago, may not support him at the present time with all his increase in desires (page 106).

It is no solution of any question merely to put other families back on disused farms. It is worse than no solution to place there a more ignorant family than was on the place originally; and yet there is a movement all over the country to place raw foreigners on such farms as owners or renters. Because these farms are cheap, they appeal to city people, and they become temptations to real estate dealers. Bargain-counter farms are rarely good investments.

What is to be done with these farms is, at bottom, a plain economic question. If they will not pay in ordinary farming, no one should be forced to occupy them. They might be well utilized, in many cases, for community or county forestry purposes. Every county in the East that has many remote and difficult hill lands could probably profit by a system of public forestry, organized on a comprehensive state plan.

I have said that farms are abandoned for all kinds of reasons. It does not follow because a family has given up a certain farm that the place has ever been really tested on its merits; the man may not have been a farmer at all, but only a resident. Misfortune in the family, or the lack of children, may be the reason for the desertion. So it happens that some so-called abandoned farms are first-class properties to purchase as ordinary farms.

The best lands will naturally be the first to be taken up by persons who know. And the value of land for farming will depend very much on its accessibility and nearness to market. Even though it is possible to raise two hundred and fifty bushels of potatoes on a distant hilltop, it does not follow that it is profitable to raise them there. Many persons who are now living on difficult lands, would undoubtedly be much better off if they were in cities or towns; but as a rule, a man cannot safely enter a new business after forty years of age.

We must, of course, do the best we can to help the man who actually lives on one of these difficult farms, to enable him to make the very most of his opportunities. This is being done through many agencies. He has been taught in methods of soil handling, fertilizing, grass-growing, stock-raising, drainage, and many other particular features. But it is also important that we do not encourage others to enter the same condition.

So I have no fear of the abandoned farm, although I wish that we had a fundamental treatment of the whole situation,—like state programs,—so that lands in the process of returning to nature may be managed in a large and systematic way, that they might contribute the best results to the community and the country. We now know how to make these lands productive, but there is a larger question than this. Such lands—once farmed and now going fallow—may be found from

California to Maine. In many cases they are not being abandoned rapidly enough, and this accounts for the human tragedy connected with some of the old homesteads. But they will all be used in good time, and we shall need them.

Little of the older country is worn out. Some of the best land values now lie in the old East and South. Movement to these lands from the Western lands is now beginning, and this is a sound tendency, as are most spontaneous movements inside the farming business itself; the railroads and real estate dealers may be expected to even up the situation.

The new farming.

Although the ratio of farmers to the whole population may still decrease, the actual number of farmers will increase. The rural districts will fill up. More young men and women will remain on farms and more persons will go from towns to farms as rapidly as the business becomes as lucrative as other businesses requiring equal investment, risks, and intelligence. The open country will probably fill up mostly with the natural increase of the country population, and there will be some to spare for the cities. We shall face the question of congestion of farm districts.

The general growth of population will make additional demands on the farm, not only because there will be more persons to supply, but also because desires increase with the increase of wealth. It may require no more food to sustain a well-to-do person than a poor-to-do person, but as one increases his income he greatly extends the range of his food and improves its quality. Luxuries increase.

But beyond his actual food, one's desires increase directly with his income; and, aside from the minerals and metals, most of the material that is used in the arts and manufactures, in clothing, shelter, and adornment, is raised from the land. The human-food products do not comprise one-half the output of the land.

We have covered in a way the "easy" farming regions. But in the end, all the country will be needed for productive uses; and the best civilization will come only when we conquer the difficult places as well as utilize the easy ones. We shall develop greater skill in farming than we have yet dreamed of. The raw and ragged open country that we see everywhere from trolley-lines and railway-trains is not at all a necessary condition; it is only a phase of a transition period between the original conquest of the country and the growing utilization of our resources. The more completely we conquer and utilize it, the more resourceful and hopeful our people should be. Country life will become more differentiated and complex. Speaking broadly, we are now in the rough and crude stage of our agricultural development; but the situation will develop only as it pays and satisfies persons to live in the country.

To meet the economic, social, educational, religious, and other needs of these great open regions will require the very best efforts that our people can put forth; and our institutions are not now sufficiently developed to meet the situation adequately.

RECLAMATION IN RELATION TO COUNTRY LIFE; AND THE RESERVE LANDS

ALL forms of reclamation, by which lands are made available for agricultural use, profoundly affect society and institutions; and any person who is interested in rural civilization must necessarily, therefore, be interested in these means and their results. Because reclamation by irrigation has progressed farther than other means, and has become a national policy, I shall confine my remarks to it chiefly.

The best rural civilization will develop out of native rural conditions rather than be imposed from without. Irrigation makes a rural condition: it provides the possibility for a community to develop; and it must, therefore, color the entire life of the community.

Irrigation communities are compact. As all the people depend on a single utility, so must the community life tend to be solidified and tense. Probably no other rural communities will be so unified and so intent on local social problems. We shall look, therefore, for a very distinct and definite welfare to arise in these communities; and they will make a peculiar contribution to rural civilization.

The life of the irrigation community will be expressed not only in institutions of its own, but in a literature of its own. Much of the world's literature does not have significance to country-life conditions, and very little of it has significance to an irrigation civilization. I look for poetry to come directly out of the irrigation ditch and to express the outlook of the people who depend for their existence on the canal and the flood-gate.

The interests of society in the work.

The people have made it possible for irrigation-reclamation to be developed; for whether the work is performed by government directly or by private enterprise, it nevertheless rests mostly on national legislation; and this legislation expresses the consent and the interest of society in the work. All the people have not only a right to an interest in irrigation-reclamation, but they carry an obligation to be interested in it, since it reclaims and utilizes the fundamental heritage of all the people. I take it that society's interest in the work is of two kinds: to see that the land is properly utilized and protected; to see that persons desiring homes shall have an opportunity to secure them. Society is not interested in speculation in land or in mere exploitation.

I hope that the irrigation people realize their obligation to the society that makes it possible for them to develop their irrigation systems. Not every person in the nation agrees to the policy of national reclamation, but society has given it a trial. The people in the West are interested in developing their localities and their commonwealths, and in securing settlers to them; and with this feeling we all must sympathize. The people in the

East have a remoter interest, but it is none the less real. I have no fear that the irrigation-settlement of the West will set up disastrous competition in products with the East, as many Eastern people anticipate; the areas involved in the new irrigation projects are too small and the development too slow for that. But there is danger that the producing-power of the land may not be safeguarded, and all the people, East as well as West, must have concern for use of Western land. The very fact that irrigation-farming is intensive increases the danger. From an agricultural point of view, the greatest weakness in this farming is the fact that the animal, or live-stock, does not occupy a large place in the system. Other systems of maintaining fertility must be developed.

Society has a right to ask that we be careful of our irrigated valleys. They are abounding in riches. It is easy to harvest this wealth, by the simple magic of water. We will be tempted to waste these riches, and the time will come quickly when we will be conscious of their decline. This seems remote now, but the danger is real. Not even the fertility of the irrigation waters will maintain the land in the face of poor agricultural practice.

I am not contending that irrigation-farming is proceeding in a wasteful way, or that systems are not developing that will protect society; I am calling attention to the danger and to the interest of all the people in this danger; and I hope that we may profit by the errors of all new settlements thus far made in the history of the world.

It is the flat valleys of the great arid West that will be opened by irrigation. These valleys are small areas compared with the uplands, the hills, and the unirrigable regions. Society is interested also that we be careful of the uplands and hills, for in the arid regions they give small yield in forage and in timber; this forage and timber must be most thoughtfully protected. When the producing-power of the irrigated lands begins to decline, the West cannot fall back on its dry hills.

Reclamation and Country Life

We are everywhere in need of better agriculture, not only that every agriculturist may do a better business, but also that agriculture may contribute its full share to the making of a better civilization. Here and there, as we learn how to adapt ourselves to the order of nature, we begin to see a really good agriculture in the process of making. A good agriculture is one that is self-sustaining and self-perpetuating, not only increasing its yields year after year from the same land, but leaving the land better and richer at each generation. This must come to pass from the land itself and from the animals and crops that one naturally brings to the land, and not merely by the addition of mined fertilizing materials brought from the ends of the earth. Thus far in history, it is only when the virgin fatness begins to be used up, speaking broadly, that we put our wits to work. Then the rebound comes. The best agriculture thus far has developed only after we have struck bottom, and we begin a constructive effort rather than an exploitative effort; and this comes in a mature country. This is why so great part of the European agriculture is so much better than our own, and why in old New England such expert and hopeful farming is now beginning to appear. The East is in the epoch of rebound. The East is in the process of becoming more fertile; the West is in the process of becoming less fertile.

In Western North America, the business systems have been developed to great perfection, and the people are possessed of much activity, and are so far escaped from tradition that they are able to do things in new ways and to work together. I hope that this great region also will apply at the outset all the resources of business and of science to develop an agriculture that will propagate itself.

A broad reclamation movement.

When all the lands are taken that can be developed or reclaimed by private resources, there remain vast areas that require the larger powers, and perhaps even the larger funds, of society (or the government) to bring into utilization. One class of lands can be utilized by means of irrigation. This form of land-reclamation is much in the public mind, and great progress has been made in it.

There remain, however, other lands to be reclaimed by other means. There is much more land to be reclaimed by the removal of water than by the addition of water. There are many more acres to be adapted to productive uses by forest planting and conservation than by irrigation. There are vastly larger areas waiting reclamation by the so-called "dry-farming" (that is, by moisture-saving farming completely adapted to dry regions). And all the land in all the states must be reclaimed by better farming. I am making these statements in no disparagement of irrigation, but in order to indicate the relation of irrigation to what should be a recognized national reclamation movement.

Supplemental irrigation.

Let me say further that irrigation is properly not a practice of arid countries alone.

Irrigation is for two purposes: to reclaim land and make it usable; to mitigate the drought in rainfall regions. As yet the popular imagination runs only to reclamation-irrigation. This form of irrigation is properly regulated by the federal government.

Now and then a forehanded farmer in the humid region, growing high-class crops, installs an irrigation plant to carry him through the dry spells. As our agriculture becomes more developed, we shall greatly extend this practice. We shall find that even in humid countries we cannot afford to lose the rainfall from hills and in floods, and we shall hold at least some of it against the time of drought as well as for cities and for power. We have not yet learned how to irrigate in humid regions, but we certainly shall apply water as well as manures to supplement the usual agricultural practices.

We must learn to reckon with drought as completely as we reckon with winter or with lessening productiveness. We probably lose far more from dry spells than from all the bugs and pests.

We need reserves.

But even though we should recognize a national reclamation movement to include all these phases and others, it may not be necessary or advisable in the interest of all the people, that every last acre in the national domain be opened for exploitation or settlement in this decade or even in this century. The nation may well have untouched reserves. No one knows what our necessities will be a hundred years hence. Land that has never been despoiled will be immeasurably more valuable to society then than now; and society holds the larger interest.

When the pressure of population comes, we shall fall back on our reserves. The rain-belt states will fall back on their wet lands, their uplands, and their hills. These hills are much more usable than those of the arid and semi-arid West can ever be. The Eastern and old Southern states have immense reserves, even though the titles may be largely in private ownership. New York is still nearly half in woods and swamps and waste, but practically all of it is usable. New York is an undeveloped country, agriculturally. The same is true of New England and Pennsylvania and great regions southwards. Forests and sward grow profusely to the summits of the mountains and hills. Vast areas eastward are undeveloped and unexploited. Even the regions of the so-called "abandoned farms" are yet practically untouched of their potential wealth.

I have no regret that these countries are still unsettled. There is no need of haste. When the great arid West has brought every one of its available acres into irrigation, and when population increases, the Eastern quarter of the country will take up the slack. It is by no means inconceivable that at that time the Eastern lands, newly awakened from the sleep of a century, will be the fresh lands, and the older regions will again become the new.

We should be careful not to repeat, even on a small scale, the recklessness and haste with which we have disposed of our reserves before their time.

WHAT IS TO BE THE OUTCOME OF OUR INDUSTRIAL CIVILIZATION?

WE know that the whole basis of civilization is changing. Industry of every kind is taking the place of the older order. Its most significant note is that it brings the people of the world together in consultation and in trade. We are escaping our localism, and we look on all problems in their relation to all mankind. Brotherhood has become a real power in the world.

But what does industry in itself, including all forms of land-culture, offer as an ultimate goal to civilized man? What are to be the man's ideals toward which he should lead his thoughts?

I am not one of those who consider a sordid and commercial end to be the necessary result of industrialism. We must develop the ideals in an industrial civilization, that they may lead us into the highest personal endeavor; and everywhere it should be possible for a man to make the most of himself. There must be something in every business beyond the financial gain if it is to make any final contribution to civilization. Finding this ultimate, industrial society will grow into perfect flower.

So far as agriculture is concerned, I see two points of high endeavor within the business, lying beyond the making of a good living, and toward which the coming countryman may set his imagination.

(1) *The making of a new society.*

A new social order must be evolved in the open country, and every farmer of the new time must lend a strong hand to produce it. We have been training our youth merely to be better farmers; this, of course, is the first thing to do, but the man is only half trained when this is done. What to do with the school, the church, the rural organizations, the combinations of trade, the highways, the architecture, the library, the beauty of the landscape, the country store, the rousing of a fine community helpfulness to take the place of the old selfish individualism, and a hundred other activities, is enough to fire the imagination and to strengthen the arm of any young man or woman.

The farmer is to contribute his share to the evolution of an industrial democracy.

(2) *The fighting edge.*

Theodore Roosevelt, with his power to discern essentials, has given us a good rallying phrase in "the fighting edge." When man ceases to be a conqueror, he will lose his virility and begin to retrograde. As localism gives way to brotherhood, militarism will pass out; but this does not mean that mankind will cease to contend.

The best example I have seen of the development of determination and fine social brotherhood is in the making of the Panama Canal. The making of the Canal is in every sense a conquest. It is a new civilization that the 40,000 or 50,000 folk are constructing down there, and every man, whether he is employed in the commissariat, the sanitary department, in an office, on a steam shovel, or with a construction gang, will tell you that he is building the Canal. All these people are giving a good account of themselves because they are doing the work under the flag and because they are contending with vast difficulties.

We have scarcely begun even the physical conquest of the earth. It is not yet all explored. The earth is an island, and it is only two years ago that we got to one end of it. There are mountains to pierce, sea-shores to reclaim, vast stretches of submerged land to drain, millions of acres to irrigate and many more millions to utilize by dry-farming, rivers to canalize, the whole open country to organize and subdue by means of local engineering work, and a thousand other great pieces of construction to accomplish, all calling for the finest spirit of conquest and all contributing to the training of men and women. There is no necessity that the race become flabby.

Now, my point is that the prime high endeavor laid before every farmer is to conquer his farm, and this means contest with storm and flood and frost, with blight and bug and pest, and with all the other barriers that nature has put against the man that tills the land. We have made a tremendous mistake, in my estimation, in trying to portray farming merely as an easy business. The sulky-plow has been too much emphasized. We are giving the young men more means and tools by which to wage the contest, but the contest can never stop. In the nature of things, farming cannot be an easy and simple business, and this is why it has produced a virile lot of men and women, and why it will continue to do so. It is a question whether, if our civilization is ever evened up, we shall not look again to the open country for strong working classes, for the course of much of our city industrialism is to make dependent men and managed men, and we need to exercise every precaution that it does not make clock-watchers and irresponsible gang-servers (page 139).

Farming will attract folk with the feeling of mastery in them, even more in the future than in the past, because the hopelessness, blind resignation, and fatalism will be taken out of it. Those who are not masterful cannot conquer a farm. The man weighing one hundred and fifty pounds who is afraid of a San José bug would better go to the city, where he can find some one to help him fight his battles. The farmer will learn how to adapt his scheme to nature, and how to conquer the things that are conquerable; and this should make it worth his while to be a farmer.

Needs revision in light [of] present federal law

THE FUNDAMENTAL QUESTION IN AMERICAN COUNTRY LIFE

How to make country life what it is capable of becoming is the question before us; and while we know that the means is not single or simple, we ought to be able to pick out the first and most fundamental thing that needs now to be done.

It is perfectly apparent that the fundamental need is to place effectively educated men and women into the open country. All else depends on this. No formal means can be of any permanent avail until men and women of vision and with trained minds are at hand to work out the plans in an orderly way.

And yet it is frequently said that the first necessity is to provide more income for the farmer; but this is the result of a process, not the beginning of it. And again it is said that organization is the first necessity, even to make it possible to use the education. If organization is necessary to make the best use of education, then it assumes education as its basis. Educated men will make organization possible and effective, but economic organization will not insure education except remotely, as it becomes a means of consolidating an unorganic society.

But there is no longer any need to emphasize the value of education. It would now be difficult to find an American farmer who requires convincing on this point. Yet I have desired to say that there is no other agency, using education in its broad sense, that can by any possibility be placed ahead of it.

Agriculture in the public schools.

Agriculture is now a school subject. It is recognized to be such by state syllabi, in the minds of the people, and in the minds of most school men. It is finding its way into high-schools and other schools here and there.

There is no longer much need to propagate the idea that agriculture is a school subject.

It is now our part to define the subject, organize it, and actually to place it in the schools.

We must understand that the introduction of agriculture into the schools is not a concession to farming or to farmers. It is a school subject by right.

It is the obligation of a school to do more than merely to train the minds of its students. The school cannot escape its social responsibilities; it carries these obligations from the very fact that it is a school supported by public money.

The schools, if they are to be really effective, must represent the civilization of their time and place. This does not mean that every school is to introduce all the subjects that engage men's attention, or that are capable of being put into educational form; it means that it must express the main activities, progress, and outlook of its people. Agriculture is not a technical profession or merely an industry, but a civilization. It is concerned not only with the production of materials, but with the distribution and selling of them, and with the making of homes directly on the land that produces the material. There cannot be effective homes without the development of a social structure.

Agriculture therefore becomes naturally a part of a public-school system when the system meets its obligation. It is introduced into the schools for the good of the schools themselves. It needs no apology and no justification; but it may need explanation in order that the people may understand the situation.

If agriculture represents a civilization, then the home-making phase of country life is as important as the field farming phase (page 93). As is the home, so is the farm; and as is the farm, so is the home. Some of the subjects that are usually included under the current name of home economics, therefore, are by right as much a part of school work as any other subjects; they will be a part of city schools as much as of country schools if the city schools meet their obligations. They are not to be introduced merely as concessions to women or only as a means of satisfying popular demand; they are not to be tolerated: they are essential to a public-school program.

The American contribution.

The American college-of-agriculture phase of education is now well established. It is the most highly developed agricultural education in the world. It is founded on the democratic principle that the man who actually tills the soil must be reached,—an idea that may not obtain in other countries.

We are now attempting to extend this democratic education by means of agriculture to all ages of our people, and there is promise that we shall go farther in this process than any people has yet gone; and this fact, together with the absence of a peasantry, with the right of personal land-holding, and with a voice in the affairs of government, should give to the people of the United States the best country life that has yet been produced.

America's contribution to the country-life situation is a new purpose and method in education, which is larger and freer than anything that has yet been developed elsewhere, and which it is difficult for the Old World fully to comprehend.

The founding of the great line of public-maintained colleges and experiment stations means the application of science to the reconstruction of a society; and it is probably destined to be the most extensive and important application of the scientific method to social problems that is now anywhere under way.

The dangers in the situation.

It is not to extol our education experiment that I am making this discussion, but to measure the situation; and I think that there are perils ahead of us, which we should now recognize.

There are two grave dangers in the organization of the present situation: (1) the danger that we shall not develop a harmonious plan, and thereby shall introduce competition rather than coöperation between agencies; (2) the danger that the newer agencies will not profit fully by our long experience in agriculture-teaching.

An internal danger is the giving of instruction in colleges of agriculture that is not founded on good preparation of the student or is not organized on a sound educational basis. Winter-course and special students may be admitted, and extension work must be done; but the first responsibility of a college of agriculture is to give a good educational course: it deals with education rather than with agriculture, and its success in the end will depend on the reputation it makes with school men.

There is also danger that new institutions will begin their extension work in advance of their academic educational work; whereas, extension and propaganda can really succeed only when there is a good background of real accomplishment at home.

There is necessity that we now reorganize much of our peripatetic teaching. It is no longer sufficient to call persons together and exhort them and talk to them. We have come about to the end of agricultural propaganda. All field and itinerant effort should have a follow-up system with the purpose to set every man to work on his own place with problems that will test him. We have been testing soils and crops and fertilizers and live-stock and machines: it is now time to test the man.

There is also danger that we consolidate too many rural schools in towns. If it is true that the best country life is developed when persons live actually on their farms, then we should be cautious of all movements that tend to centralize their interests too far from home, and particularly to centralize them in a town or in a village. The good things should come to the farm rather than that the farm should be obliged to go to the good things.

The present educational institutions.

We must first understand what our institutions of education are. The extension of agriculture-education in institutions in the United States (beyond the regular colleges of agriculture) is in four lines: as a part of the regular public-school work; in unattached schools of agriculture publicly maintained; in departments attached to other colleges or uni-

Fundamental Question 69

versities; in private schools. The last category (the private schools) may be eliminated from the present discussion.

The separate or special-school method is well worked out in Wisconsin (county plan), in Alabama and Georgia (congressional-district plan), Minnesota (regional plan), with other adaptations in Louisiana, Oklahoma, Michigan, Maryland, and elsewhere.

In New York, the movement for special schools has taken an entirely new direction. Two schools are connected with existing institutions of higher learning of long-established reputation (being the only schools of this kind, state-maintained, attached to liberal arts universities) and one is unattached; none of them has a defined region or territory. These institutions are established on a more liberal financial plan than the special schools of other states, standing somewhat between those schools and the agricultural college type.

While much publicity has been given to the unattached-school plan, the main movement is the adding of agriculture-education to the existing public-school systems. Only eight or ten of the states have entered into any regular development of separate or unattached schools, whereas in every state the movement for agriculture in the public schools is well under way. The public schools are of definite plan; the unattached schools are of several plans, or of no plan; and in some states an intermediate course is developing by the establishing of public high-schools (one to a county, a congressional district, or other region) in which instruction in agriculture and household subjects is highly perfected.

Aside from the foregoing particular institutions, many general colleges and universities are introducing agricultural work in order to meet the increasing demand and to keep up with educational progress.

Agricultural work is proceeding in nearly all the states under the auspices of the United States Department of Agriculture, some of it distinctly educational in character; and there is ~~agitation for the passage of~~ a national bill to further secondary or special agriculture-education in the states.

State departments of agriculture, the indispensable experiment stations, veterinary colleges, departments of public instruction, farmers' institutes, voluntary societies, are all attacking the country-life problem in their own ways; and the powerful work of the agricultural press, although not coming within the scope of this paper, should not be overlooked as an educational agency.

In the meantime, the colleges of agriculture are growing rapidly and are approaching the subject from every side, and are assuming natural and inevitable leadership.

The need of plans to coördinate this educational work.

There is no doubt that all these agencies are contributing greatly to the solution of the rural problem, and there is now probably very little inharmony and little duplication of effort. In the newness and enthusiasm of the effort, good fellowship holds the work together in all the states or at least keeps it from collision. But the situation is inherently weak, because there is no plan or system, and no united discussion of the grounds on which the work rests. I have been in correspondence on this question with public men in every state in the Union, and I find a general feeling that the present situation is fraught with danger, and that there is great need of organization or at least of federation of the forces within each state; and ultimately there must be federation on a national basis. The work should be coöperative rather than competitive.

What is to be the policy of the state in agriculture-education? Where is the headship to lie? What are to be the spheres of the different institutions and agencies? What board or agency is to correlate and unify all the parts, to insure a progressive and well-proportioned program?

Outline of a state plan.

A general law should define the state's policy in education by means of agriculture and in the development of rural affairs, and outline methods that it proposes to follow, so that the work may be coördinated throughout the state and that a definite plan may be projected.

The duties of all the classes of institutions should be defined and relations should be established between them. The people should know to what they are committing themselves.

This law should not, of course, be designed to suppress the activities of any institution. It might not place any institution under the domination of any other institution. The schools, colleges, and other institutions for the betterment of agriculture should have their own autonomy and responsibility, and they should be developed to the highest point of efficiency in their respective spheres.

The fundamental consideration in such a law should be to develop the agriculture and advance the country life of the state by organizing the work of all the agencies on a systematic plan, so that an orderly development may be secured. Such a recognized general policy should do much to insure each institution in the system its proper state support.

It is probably too much to expect that a fundamental state law could be projected abstractly. Laws are gradually built up to meet urgent needs as they arise; but if the principles are kept in mind, the making of separate and special laws might be so guided as to produce a harmonious result.

Some of the particular points that I think should be desired in such a law or series of laws are these:

1. It should propound a policy in the development of country life;
2. It should name the classes of institutions that it proposes to utilize in the execution of this policy;
3. It should define the functions of the different classes of institutions;
4. It should state the organic relationships that ought to exist between them all;
5. It might provide an advisory council to guide agricultural education and advancement in the state. I think that the directors or responsible heads of such institutions established for the betterment of agriculture throughout the state should constitute such consulting body, to which questions of

policy and procedure should be referred and which, of course, should serve without remuneration. This council might include also the commissioner of agriculture and the superintendent of public instruction. It might be well to have one, two, or three other persons appointed by the governor. The council would constitute a natural conference of the parties that are immediately responsible for this work, without taking the management of any institution out of the hands of an existing board. The idea of such a body is to further the coördination by conference, rather than to have plenary power. Its moral influence ought to be all the greater because of its lack of conferred power.

A state extension program.

As soon as a state has produced strong institutions for research and education in agriculture, it will need to provide an agency for utilizing the results. A state extension program, on a coördinating plan between all the institutions but proceeding from one educational center, and which all the institutions would have a right to use for the spread of their work among the people, could accomplish vast benefits. It should comprise institutes, utilize the state system of fairs educationally, afford an organ for the making of agricultural surveys and demonstrations, spread an educational propaganda on the agricultural law, collect and collate the experience of the farmers of the state. It could assort and apply the information that the state, at great expense, accumulates through its various separate agencies. It could utilize the students, whom the state provides with free tuition. The germ of such an enterprise already exists in most of the states.

Special local schools for agriculture.[1]

I am committed to the idea that there should be strong local centers of interest in rural com-

---

[1] See "The State and the Farmer," p. 150; "The Training of Farmers," p. 167; "Cyclopedia of American Agriculture," IV, p. 474.

munities, for thereby we develop local pride and incentive. There are several ways, on the educational side, of developing local institutions and interest.

The first way is to make it possible and practicable for the existing public schools to introduce agriculture and domestic economy. I suggest that many or most localities would do better to develop the country-life work in the existing schools than to ask the legislature for a separate special school. We have only begun to understand what such redirected and expanded schools may accomplish.

Another means of securing local knowledge and developing local interest is by the establishing of demonstration farms and field-laboratories. It is doubtful whether a permanent demonstration farm in a community is desirable; in general, the demonstration may be temporary, depending on the presence in the community of some special difficulty. In some circumstances, the enterprise may amount to a local testing station. Enterprises of this sort are bound to take on great importance in the redirection of country life.

Local societies and organizations may be encouraged to take up educational and experiment work.

Departments of agriculture will probably be added by colleges or other educational institutions, and these will serve as local centers at the same time that they reach the larger field.

Again, a winter school or short-course of, say, a month's or two months' duration may be held in different parts of the state. The localities should coöperate in the expenses, thereby becoming partakers in the enterprise.

Eventually there should be an agricultural agent resident in every county, and perhaps even for smaller regions, whose office should be to give advice, to keep track of animal and plant diseases and pests and secure the services of experts in their control, to organize conferences, winter-courses, and the like, and otherwise to be to the agricultural affairs what the pastor is to religious affairs and the teacher to educational affairs. (See "The Training of Farmers," p. 257.)

Finally, we may ask the state to place a

special school of agriculture in the locality, but only after it is clear that other means cannot produce the desired results. An unattached school of agriculture is not an easy thing to administer successfully, even at the best; and the difficulty would be all the greater if its care were to be confined to local boards, which would probably have small understanding of the peculiar educational requirements. It is probable that a state may wisely establish a very few special schools, but an educational program needs first to be worked out, a competent system of control must be found, and the people should know in advance what is involved. It is not enough merely that a locality desires a school: the larger question is the state's interest. In all local enterprises of this kind in which state aid is asked for, it ought to be understood that the locality itself is to coöperate in the securing of equipment and funds.

The lessons of experience.

The demand for agriculture-education is now widespread; the subject is becoming "popular." All kinds of plans are being tried or discussed.

Persons do not seem to realize that we have had about one hundred years of experience in the United States in agriculture-education, and that this experience ought to point the way to success, or at least to the avoiding of serious errors. The agricultural colleges have come up through a long and difficult route, and their present success is not accidental, nor is it easy to duplicate or imitate. First and last, about every conceivable plan has been tried by them, or by others in their time or preceding them; and this experience ought to be utilized by the other institutions that are now being projected in all parts of the country.

Plans that certainly cannot succeed are now being projected. The projectors seem to proceed on the idea that it requires no background of experience to enable an institution to teach agriculture, whereas agriculture-education is the most difficult and also the most expensive of all education yet undertaken.

To teach agriculture merely by giving a new

direction or vocabulary to botany, chemistry, geology, physics, and the like is not to teach agriculture at all, although it may greatly improve these subjects themselves. To put a school of agriculture in the hands of some good science-teacher in a general college faculty with the idea that he can cover the agricultural work and at the same time keep up his own department, is wholly ineffective (except temporarily) and out of character with the demands of the twentieth century (but in high-schools a good science teacher may handle the work, or an agriculture teacher may carry the science). To suppose that "agriculture" is one subject for a college course, to be sufficiently represented by a "chair," is to miss the point of modern progress. To give only laboratory and recitation courses may be better than nothing, but land-teaching, either as a part of the institution or on adjacent farms, must be incorporated with the customary school work if the best results are to be secured. To make a school farm pay for itself and for the school is impossible unless the school is a very poor or exceedingly small one; and yet this old fallacy is alive at the present day. To have a distant farm to visit and look at, in order to "apply" the "teachings" of chemistry, botany, and the like, falls far short of real agriculture instruction. To develop a "model farm" that shall be a pattern to the multitude in exact farming is an exploded notion: there are many farmers' farms that are better adapted to such purpose (the demonstration farm is the modern adaptation of the idea, and it is educationally sound).

To teach agriculture of college grade requires not only persons who know the subject, but an organization well informed on the educational administration that is required. There must be a body of experience in this line of work behind any teaching on a college plane that shall be really useful; when this body of experience does not exist, the work must necessarily grow slowly and be under the most expert direction. The presumption is still against successful agriculture work in the literary and liberal arts institutions, because such teaching demands a point of view on education that the men in these institutions are likely not to possess. Agriculture cannot be introduced in the same way that a department or chair of history or mathematics can be organized; it requires a different outlook on educational procedure, a different order of equipment and of activities, and its own type of administration.

I am much afraid that some of the newer unattached institutions, in their eagerness to make departures and to be self-sufficient, will not profit by our long development, and that the secondary schools and others may make many of the mistakes that the regular colleges of agriculture long ago have made. The presumption is against any school that expects to develop merely a local enterprise, without reference to other schools or to experience.

I am sure we all want to encourage the introduction of agriculture into all educational institutions, but we should not be misled merely by the word "agriculture"; and in the interest of good work we should be careful not to encourage any enterprise of this kind until convinced that it has been well studied and that it will be administered in the interest of rural progress.

WOMAN'S CONTRIBUTION TO THE COUNTRY-LIFE MOVEMENT

On the women depend to a greater degree than we realize the nature and extent of the movement for a better country life, wholly aside from their personal influence as members of families. Farming is a co-partnership business. It is a partnership between a man and a woman. There is no other great series of occupations in which such co-partnership is so essential to success. The home is on the farm, and a part of it. The number of middle-aged unmarried men living on farms is very small. It is quite impossible to live on a farm and to run it advantageously without family relation.

It follows, then, that if the farming business is to contribute to the redirection of country life, the woman has responsibilities as well as the man. As the strength of a chain is determined by its weakest link, so will the progress of rural civilization be determined by the weakness of the farm as an enonomic unit, or by the weakness of the home as a domestic and social unit.

Now, the farmer himself cannot have great influence in redirecting the affairs of his community until he is first master of his own problem, — that is, until he is a first-class farmer. In the same way, a woman cannot expect to have much influence in furthering the affairs of her rural community until she also is master of her own problem, and her problem is primarily the home-making part of the farm. In the mastering of his or her own problem, the farmer or his wife may also contribute directly to the progress of the community. Every advance in the management of the household contributes to the general welfare: it sets new ideas under way.

If the farming business must in general be reorganized, so also must the householding part of it be reorganized. The solution of the farm-labor problem, for example, lies not alone merely in securing more farm "hands," but in so directing and shaping the business that less farm hands will be needed to secure a given economic result; so also the solution of the household-labor problem is not merely the securing of more household help, but the simplification of householding itself.

So far as possible, the labor that is necessary to do the work of the open country, whether in-doors or out-doors, should be resident labor. The labor difficulty increases with reduction in the size of the family. Families of moderate size develop responsibility, and coöperation is forced on all members of it, with marked effect on character. The single child is likely to develop selfishness rather than coöperation and sense of responsibility. To a large extent, the responsibility of the household should rest on the girls of the family; and all children, whether boys or girls, should be brought up in the home in habits of industry.

It is fairly possible by means of simplification of householding and by a coöperative industry amongst all members of the family, so to reduce the burden of the farm wife that she may have time and strength to give to the vital affairs of the community.

The affairs of the household.

It is essential that we simplify our ideals in cooking, in ornament, in apparel, and in furnishing; that we construct more convenient and workable residences; that we employ labor-saving devices for the house as well as for the barns and the fields.

We are so accustomed to the ordinary modes of living that we scarcely realize what amount of time and strength might be saved by a simplified table and by more thoughtful methods of preparing food. In respect to houses, it should be remembered that the present farm dwellings are getting old. A good part of the farm houses must soon be either rebuilt or remodeled. The first consideration is so to build or remodel them that steps may be saved to the housewife. We have not thought, in the past, that a woman's steps cost time and energy. Within twenty years all first-class farm houses will have running water, both into the house and out of the house.

It is rather strange that in our discussions of the farm-labor problem, we do not realize that a gasoline engine or a water engine may save the labor of a man. Farmers are putting power into their barns. They should also put power into the house. This may be accomplished by means of a small movable engine that can be used either in the house or barn, or else by installing an engine in a small building betwixt the house and the barn, so that it can be connected either way. This can be used to lighten much household labor, as pumping of water, meat-chopping, laundering, dish-washing, vacuum-cleaning, and the like.

Eventually, there must be some form of community coöperation in the country to save household labor. Already the care of milk has been taken from great numbers of farm homes by the neighborhood creamery, or at least by the building of a milk-house in which the men by the use of machinery perform labor that was once done by the housewife. Whenever there is a coöperative creamery, there may also be other coöperative attachments, as a laundry, or other appliances. It will be more difficult to bring about coöperation in these regards in country districts than in the city, but with the coming of good roads, telephones, and better vehicles, it will be constantly more easy to accomplish.

The affairs of the community.

I have said that it is important that the country woman have strength and time to engage in the vital affairs of the community. I am thinking of the public sentiment that women can make on any question that they care to discuss thoroughly and collectively, whether this sentiment is for better orcharding, better fowls, better roads, extending of telephones, improving the schoolhouse or church or library. It is needful that women in the country come together to discuss woman's work, and also to form intelligent opinions on farming questions in general.

The tendency of all "sociables" in country

and town is to bring persons together to eat, to gossip, and to be entertained. We need to redirect all these meetings, and to devote at least a part of every such meeting to some real and serious work which it is worth while for busy and intelligent persons to undertake.

Every organization of women should endeavor to extend its branches and its influence into the open country as well as into the cities and towns. Every public movement now has responsibility to country-life questions as well as to town questions.

I think it important that there be some means and reason for every farm woman going away from home at least once a week, and this wholly aside from going to town to trade. There should be some place where the women may come together on a different basis from that of the ordinary daily routine and the usual buying and selling. I do not know where this social center should develop, and in an atmosphere that is not conducive to gossip. In some neighborhoods it might focalize in the church parlor. The center should be permanent, if possible. It should be a place to which any woman in a community has a right to go. An ideal place for such a center would be the rural library, and I hope that such libraries may arise in every country community, not only that they may supply books but that they may help provide a meeting-place on semi-social lines. I think that if I were a woman in charge of a rural library, I should never be satisfied with my work until I had got every woman in the community in the habit of coming to the library once every week.

The woman's outlook.

The woman needs very much to have the opportunity to broaden her horizon. The farmer has lived on his farm; he is now acquiring a world outlook. The woman has lived in her house; she also is acquiring a world outlook. As the house has been smaller and more confining than the farm, it has followed that woman's outlook has been smaller than man's.

I think it is necessary also that the woman

of the farm, as well as the man, have a real anchor in her nature environment. It is as necessary to the woman as to the man that her mind be open to the facts, phenomena, and objects that are everywhere about her, as the winds and weather, the plants and birds, the fields and streams and woods. It is one of the best resources in life to be able to distinguish the songs and voices of the common fields, and it should be a part of the education of every person, and particularly of every country person, to have this respite. The making of a garden is much more than the growing of the radishes and strawberries and petunias. It is the experience in the out-of-doors, the contact with realities, the personal joy of seeing things germinate and grow and reproduce their kind.

The means of education.

If country women are to develop a conscious sense of responsibility in country-life betterment, education facilities must be afforded them. The schools must recognize home-making subjects equally with other subjects.

What becomes a part of the school eventually becomes a part of the life of the people of the region.

The leadership in such subjects is now being taken by the colleges of agriculture. This is not because domestic subjects belong in a college of agriculture more than elsewhere, but only that these colleges see the problem, and most general colleges or universities have not seen it. The college of agriculture, if it is highly developed, represents a civilization rather than a series of subjects; and it cannot omit the home-making phase if it meets its obligation to the society that it represents (page 64).

If the customary subjects in a college of agriculture are organized and designed to train a man for efficiency in country life and to develop his outlook, so also is a department of home economics to train a woman for efficiency and to develop her outlook to life.

Home economics is not one "department" or subject, in the sense in which dairying or entomology or plant-breeding is a department.

It is not a single specialty. It stands for the whole round of woman's work and place. Many technical or educational departments will grow out of it as time goes on. That is, it will be broken up into its integral parts, and it will then cease to be an administrative department of an educational institution; and very likely we shall lose the terms "home economics," "household economics," "domestic science," and the rest.

I would not limit the entrance of women into any courses in a college of agriculture; on the contrary, I want all courses open to them freely and on equal terms with men; but the subjects that are arranged under the general head of home economics are her special field and sphere. On the other hand, I do not want to limit the attendance of men in courses of home economics; in fact, I think it will be found that an increasing number of men desire to take these subjects as the work develops, and this will be best for society in general.

Furthermore, I do not conceive it to be essential that all teachers in home economic subjects shall be women; nor, on the other hand, do I think it is essential that all teachers in the other series of departments shall be men. The person who is best qualified to teach the subject should be the one who teaches it, whether man or woman.

As rapidly as colleges and universities come to represent society and to develop in all students a philosophy of life, the home-making units will of necessity take their place with other units.

HOW SHALL WE SECURE COMMUNITY LIFE IN THE OPEN COUNTRY?

It is generally agreed that one of the greatest insufficiencies in country life is its lack of organization or cohesion, both in a social and economic way. Country people are separated both because of the distances between their properties, and also because they own their land and are largely confined to its sphere of activities. There is a general absence of such common feeling as would cause them to act together unitedly and quickly on questions that concern the whole community, or on matters of public moment.

This lack of united action cannot be overcome by any single or brief process, but as one result of a general redirection of rural effort and the stimulating of a new or different point of view toward life. It will come as a result of a quickened agricultural life rather than as an effect of any direct plan or propaganda. When the rural social sense is thoroughly established, we shall be in a new epoch of rural civilization.

It is now the habit to say that this desired rural life must be coöperative. A society that is fully coöperative in all ways is one from which the present basis of competition is eliminated. I think that no one intends, however, in the common discussion of coöperation to take sides on the theoretical question as to whether society in the end will be coöperative or competitive; these persons only mean that coöperative association is often the best means to secure a given result and that such association may exert great educational influence on the coöperators.

Theoretically, the coöperative organization of society may be the better. Practically, a capitalistic organization may be better: it quickly recognizes merit and leadership; but if it is better, it is so only when it is very carefully safeguarded.

It cannot be contended that a coöperative

Community Life 99

organization is correct because the majority rules. Majorities show only what the people want, not necessarily what is best. Minorities are much more likely to be right, because thinking men and fundamental students are relatively few; yet it may be the best practice, in common affairs, to let the majority have its way, for this provides the best means of education.

It will now be interesting to try to picture to ourselves some of the particular means by which social connection in the open country may be brought about. It is commonly, but I think erroneously, thought that community life necessarily means a living together in centers or villages. I conceive, on the contrary, that it is possible to develop a very effective community mind whilst the persons still remain on their farms. In this day of rapid communication, transportation, and spread of intelligence, the necessity of mere physical contiguity has partly passed away.

That is, "isolation," as the city man conceives of it, is not necessarily a bar to community feeling. The farmer does not think in terms of compact neighborhoods, trolley cars, and picture shows. The country is not "lonely" to him, as it is to a city man. He does not search for amusement at night.

Hamlet life.

It is said that the American farmer must live in hamlets, as does the European peasant. The hamlet system that exists in parts of Europe represents the result of an historical condition. It is the product of a long line of social evolution, during which time the persons who have worked the land have been peasants, and to a greater or less extent have not owned the land that they have worked.

Some persons fear that the American farmer is drifting toward peasantry. This notion has no doubt arisen from the fact that in certain places the man who works the land is driven to great extremity of poverty, and he remains uneducated and undeveloped; but ignorance and poverty do not constitute peasantry. The peasanthood of the Old World is a social caste or class, and is in part a remnant of feudal government, of religious subjugation, and of the old necessity of protection. The present day is characterized by the rise of the people on the land; this movement is a part of the general rise of the common people (or the proletariat). If popular education, popular rights, and the general extension of means of communication signify anything, it is that we necessarily are developing away from a condition of peasantry rather than toward it, however much degradation or unsuccess there may be in certain regions or how much inadjustment there may be in the process (page 129).

In contradistinction to the exclusive hamlet system of living together, I would emphasize the necessity that a first-rate good man must live on the farm if he is to make the most of it. Farming by proxy or by any absentee method is just as inefficient and as disastrous in the long run as the doing of any other business by proxy; in fact, it is likely to be even more disastrous in the end because it usually results in the depletion of the fertility of the land, or in the using up of the capital stock; and this becomes a national disaster. I hold that it is essential that the very best kind of people live actually on the land. The business is conducted on the land. The crops are there. The live-stock is there. The machinery is there. All the investment is in the place itself. If this business is to be most effective, a good man must constantly be with it and manage it. A farm is not like a store or a factory, that is shut up at night and on Sunday.

The more difficult and complex the farming business becomes, the greater will be the necessity that a good man remain with it.

We must remember also that if the landowner or the farmer lives in a village or hamlet and another man lives on his farm, a social division at once results, and we have a stratification into two classes of society; and this works directly against any community of interest. It is not likely that the farmer who has retired to town and the hired man who works his farm under orders will develop any very close personal relation. The farmer becomes an extraneous element injected into the town, and has little interest in its welfare, and he has taken his personality, enterprise, and influence out of the country. He is in a very real sense "a man without a country." The increase of his living expenses in town is likely to cause him to raise the rent on his farm, or, if the tenant works for wages, to reduce the improvements on the place to the lowest extent compatible with profit. We need above all things to produce such a rural condition as will satisfy the farmer to live permanently in the country rather than to move to town when the farm has given him a competence.

I am not to be understood as saying that farmers ought never to live in town. There will always be shifting both ways between town and country. In some cases, small-area farming develops around a village; or a village grows up because the farms are small and are intensively handled. In irrigation regions, the whole community may be practically a hamlet or village. In parts of the Eastern states, small farmers sometimes live in the village and go to the farm each day, to work it themselves.

But all these are special adaptations, and do not constitute a broad agricultural system.

In time we probably shall develop a new kind of rural settlement, one that will be the result of coöperative units or organizations, and not a consolidation about the present kinds of business places; but it is a question whether these will be villages or hamlets in the sense in which we now use these words.

The category of agencies.

My position, therefore, is that we must evolve our social rural community directly from the land itself, and mostly by means of the resident forces that now are there.

This being our proposition, it is then necessary to discover whether, given permanent residence on pieces of land, it is still possible to develop anything like a community sense. I do not now propose to discuss this question at any length, but merely to call attention to a few ways in which I think the neighborhood life of the open country may be very distinctly improved.

In this discussion, I purposely omit reference to public utilities and governmental action, because they are outside my present range. The farmer will share with all the people any needful improvement that may be made in regulation of transportation and transportation rates, in control of corporations, in equalizing of taxation, in providing new means of credit, in extending means of communication, in revising tariffs, in reforming the currency, and in perfecting the mail service.

To work out the means of neighborhood coöperation, there should be sufficient and attractive meeting places. The rural schoolhouse is seldom adapted to this purpose. The Grange hall may not represent all the people. The church is not a public institution. Libraries are yet insufficient. Town halls are few, and usually as unattractive as possible. There is now considerable discussion of community halls. Several of them have been built in different parts of the country to meet the new needs, and the practice should grow.

1. The mere *increase of population* will nec-

essarily bring people closer together, and by that much it will tend to social solidarity.

2. The natural *dividing up of large farms*, which is coming both as a result of the extension of population and from the failure of certain very large estates to be profitable, will also bring country people closer together. The so-called " bonanza farms " are unwieldy and ineffective economic units; and many farmers are " land poor."

3. We shall also *assemble farms*. The increasing population on the land will not always result in smaller farms. Most of the richer and more profitable lands will gradually be divided because, with our increased knowledge and skill, persons can make a living from smaller areas. The remoter and less productive lands will naturally be combined into larger farm areas, however, because a large proportion of such lands cannot make a sufficient profit, when divided into ordinary farm areas, to support and educate a present-day family (page 38). Contiguous areas of the better lands will be combined with them, in

order to make a good business unit. As several farms come together under one general ownership, this owner will naturally gather about him a considerable population to work his lands.

The probability is that, under thoroughly skillful single management, a given area of remote or low-productive lands will sustain a larger population than they are now able to sustain under the many indifferent or incompetent ownerships. It is to be hoped that some of these amalgamated areas will develop a share-working or associative farming of a kind that is now practically unknown.

4. The *re-creative life* of the country community greatly needs to be stimulated. Not only games and recreation days need to be encouraged, but the spirit of release from continuous and deadening toil must be encouraged. The country population needs to be livened up. This will come about through the extension of education and the work of ministers, teachers, and organizations. All persons can come together on a recreation basis (pp. 173, 211).

The good farmer will have one day a week for recreation, vacation, and study.

5. *Local politics* ought to further the entire neighborhood life, rather than to divide the community into hostile camps. All movements, as direct nominations, that stimulate local initiative and develop the sense of responsibility in the people will help toward this end.

6. *Rural government* is commonly ineffective. It needs awakening by men and women who have arrived at some degree of mastery over their conditions. We talk much of the need of improving municipal government, but very little about rural government; yet government in rural communities is inert and dead, as compared with what it might be, and there is probably as much machine politics in it, in proportion to the opportunities, as in city government. Very much of the lack of gumption in the open country is due to the want of a perfectly free and able administration of the public affairs.[1]

---

[1] See "The Training of Farmers," pp. 26-28, and "The State and the Farmer," p. 125.

The whole political organization of rural communities needs new attention, and perhaps radical overhauling. As I write these sentences, I have before me a newspaper in which a progressive surgeon expresses his opinion (which he has verified for me) on the question of supervision of health in a rural county in an Eastern state. He found the statistics too inaccurate and too indefinite to enable him to draw exact conclusions, but these are approximately the facts:

"No township seems to have deliberately paid its health officer, and but one town deliberately paid its poor physician. The others paid various bills for 'quarantine' and 'fumigating' and 'fees' and other misleading items. There was no way in which to distinguish between the care of the poor and the sick-poor except to guess and to figure on what I happened to know about. A——, the richest and largest township, has no health officers, and spent $200 for the poor in a population of 4000 people living in an area of 93 square miles. B——, the poorest township, with a population of 1000, and an area of 36 square miles, paid her health officer $28 and her poor physician $23.

"One township has 2170 inhabitants living in 51 square miles of territory, worth one and one-eighth million dollars. Its supervisor is paid $352.95 a year for a few days' work; its officers are paid $612.95. It costs $274.79 each year to elect these officers, and I understand each township is to spend about $5000 for good roads. The health officer that cares for these 2000 people over 51 miles of territory gets $42.53 a year, and the poor physician $34; while the sick-poor get helped to the munificent sum of $59.36, or two and one-half cents from each citizen. The health officers get almost exactly two cents a head for caring for the inhabitants over 51 square miles of land. The supervisor gets out of each inhabitant seventeen cents a year, the officers get thirty cents, while the sick-poor take from each citizen almost three cents. The discrepancy is too glaring to need comment. A community assessed a million dollars and probably worth two millions spends $40 a year on public health, and $60 a year on one-sixteenth of its population for sickness."

The physician proposes a county commission to take the place of the board of supervisors. He declares that the members of the board have outgrown their usefulness. "They should be junked along with other stage-coaches and a nice, new 60 h. p. county commission put in their place. The fact is that the system is wrong. Our 'government' is a survival of early times, and our science is up to date. They do not fit. You cannot expect supervisors who were useful in the time of Adam, when there were no cities, no problems, no roads, to serve in the twentieth century with its surgical treatment of degenerates, its germs and prophylactics, its preventive medicine and its scientific spirit. Supervisors could look after noxious plants and animals in the old days, and they could paper the court-house and eat fat dinners at the poor-house. They did fairly well at settling line fences, drinking sweet cider, and blarneying with insurgents.

But they are out of place when it is a question of constructing roads of macadam, of building a tuberculosis hospital for an $18,000,000 county, and especially they are out of place when it is a question of dozens of defectives in the jails and thousands outside who ought to be in hospitals."

7. A *community program for health*[1] is much needed. The farmer lives by himself in his own house, on his own place. If a disease arises in his neighbor's family, it is not likely to spread to his family. Therefore, disease has seemed to him to be a personal rather than a neighborhood matter. There is the greatest need that the farmer possess a community sense in respect to disease and sanitary conditions. If the city is the center of enlightenment, it should help the country to get hold of this problem.

We should have a thoroughgoing system of health supervision and inspection for the open country as well as in the city. Health in-

---

[1] Another discussion of rural health will be found in my "Training of Farmers," pp. 46–68. The Century Co.

spection should run out from the cities and towns into all the adjoining regions, maintaining proper connections with state departments of health. It should be continuous. It should include inspection of animals as well as of human beings. In other words, the whole region is a unit, one part depending on the other. The remarks of the physician, just quoted, indicate how great is the need of an organized health supervision for country communities.

We need meat inspection laws for meat killed and sold within the states, to supplement the inter-state law. We need community slaughter-houses in which all slaughtering of animals shall be under proper inspection. We need state milk inspection programs. It is not right that any large city should be compelled to inspect the milk throughout the state in order to protect itself. It is not right to the farming districts that such inspection should center in the city.

We must not assume that the farmer is specially guilty of sanitary faults. There are many such shortcomings in the open country, and I accept them without apology; but I can match them every one in city conditions. The fact is that the whole people has not yet risen to an appreciation of thoroughly sanitary conditions, and we cannot say that this deficiency is the special mark of any one class of our population. Persons ride along the country roads and see repulsive barn-yards, glaring manure piles, untidy back-yards, and at once make remarks about them. All these things are relegated to the rear in towns and cities and are not so visible, but they exist there.

I know that there are very filthy stables in the country districts, but I have never known worse stable conditions than I have seen in cities and towns. All progress in these directions must come slowly, and we must remember that it is expensive to rebuild and reorganize a stable. No doubt one of the reasons for the high cost of living is the demand of the people that pure-food laws shall be enacted and enforced, for this all adds to the cost of food supplies; similarly, we must expect a betterment in conditions of stabling to result in increased price of dairy products. In the cost of living we must figure the expense of having clean and pure food.

The farmer is much criticized for polluting streams; but when the farmer pollutes one stream occasionally, a city will pollute a whole system of streams continually. One of the greatest sins of society is the wholesale befoulment of streams, lakes, and water-courses. I do not see how we can expect to be called a civilized people until we have taken care of our refuse without using it to fill up ponds and lakes, and to corrupt the free water supplies of the earth.

If the countryman has been ignorant of sanitary conditions, we must remember that his ideas are largely such as he has derived from teachers, physicians, and others.

We cannot expect a man to develop within himself enough community pride and altruism to compel him to go to great expense for the benefit of the public; but he will gladly contribute his part to a public program.

8. *Local factories and industries* of whatever kind tend to develop community pride and effectiveness. Creameries have had a marked effect in this way in many places, giving the community or locality a reason for existence and a pride in itself that it never had before, or at least that it had not enjoyed since the passing out of the small factories. There is much need of local industries in the open country, whether they are distinctly agricultural or otherwise, not only for the purpose of providing additional employment for country people but to direct the flow of capital and enterprise into the country and to stimulate local interest of all kinds. It is not by any means essential that all the new life in country neighborhoods should be primarily agricultural.

Much has been said of late about the necessity of introducing the handicrafts in the open country in winter with the idea of providing work for farm people during that season. I do not look for any great extension of this idea in real agricultural sections, and for the following reasons: (1) because as better agriculture develops, the farms will of themselves employ their help more continuously. Modern diversified and intensive farming brings about this result. The present-day dairying employs men continuously. The fruit-grower needs help in winter for pruning and spraying. Livestock men need help in feeding and caring for the animals. Modern floriculture and vegetable-gardening are likely to run the year round. (2) The conditions of American country life are such that skilled handicraft has not arisen amongst the rural people, and we cannot expect that it will arise. Skilled artisanship of this kind is not the growth of a generation, nor is it a result of the utilization of merely a few weeks or months of time. (3) It is very doubtful whether such handicrafts as are often mentioned could compete in the markets with the goods produced by consolidated factories, or could find a sufficient patronage of people interested in this kind of handicraft products.

I am not arguing against the introduction of handicrafts, but wish only to call attention to what I think to be an error in some of the current discussions. I am convinced that local industries of one kind or another will find their way into the open country in the next generation, and greatly to the advantage of the country itself; but the most useful of them will be regular factories able to compete with other factories. Their largest results will come not in providing employment for persons who temporarily need it, but in developing a new community life in the places where they stand.

9. *The country store* ought to be a factor in rural betterment. How to make it so, I do not know. The country store is the nexus between the manufacturers or the city jobbers, with their "agreements," on the one hand, and the people, on the other hand, whose commercial independence the jobbers may desire to control. The country merchant takes up the cause of the large dealer, because his own welfare is involved, and he unconsciously becomes one of the agencies through which the open country is drained and restrained. The parcels post — which must come — will probably considerably modify this establishment, although

I do not look for its abolition nor desire it. Certain interests make strong opposition to the parcels post on the ground that it will ruin the country merchant and, therefore, the country town. I doubt if it will do any such thing; but even if it should, the end to be gained is not that the country merchant shall not be disturbed, but that the people at large may be benefited. No one knows just what form of readjustment the parcels post will bring about; but trade will very soon readjust itself to this condition as it has reacted to the introduction of farm machinery, good roads, the telegraph and telephone, rural free delivery.

The trader in the small town in some parts of the country is likely to own the people. He is almost necessarily opposed to coöperation and to any new movements that do not tend to enlarge his trade.

I wish we might also do something with the country hotel.

10. *The business men's organizations*, or chambers of commerce, in villages and country cities will not confine their activities within the

city boundaries in the future. A wholly new field for usefulness and for the making of personal reputation lies right here. The business organization of one village or city should extend out into the country until it meets a similar organization from the adjoining village, and the whole region should be commercially developed (pages 122–123). A chamber of commerce could exert much influence toward making a better reputation for the pack of apples, or for other output of the region.

11. The influence of certain *great corporations* is likely to be felt on the rural readjustment. This is particularly true of the new interest that railroads are taking in Eastern agriculture. A coördination between railroads and farming interests will do very much for the property of both sides; and the railroads can exercise great power in tying country communities together. The Wall Street Journal comments as follows on the situation, after calling attention to the fact that the "Eastern trunk lines have already entered upon a campaign for the encouragement of agriculture":

"Thirty-six years ago the Pennsylvania state legislature made an effort to save the farmers of that state from the damaging competition of ruinously low rates on Western grain to Eastern mills and to the seaboard. The result was practically nil. Eastern farmers were left so completely out in the cold that thousands of them sold out and went West to raise more grain there, still further to handicap the Eastern producer. The widespread bankruptcy of the middle states farmers during the eighties was a consequence partly of cut-throat competition among railroads to haul Western grain to the East at less than cost, and partly the result of a general depression from which it took ten full years to recover.

"What is it that has brought the railroads to the farmers on terms of coöperation for the development of their common territory? It is the same thing which has served the railroads so admirably in the solution of their cost problems. It is science applied to reducing the expenses of transportation in the one case, and to the greater mastery of the resources of the soil in the other case. In this lies the possibility of increasing railway freight to and from rural sources. The coöperation of transportation and agriculture, in the East especially, is not wholly new, but it is highly significant.

"Nothing could be more encouraging than the service which the railroads are beginning to render in the better distribution of population over the land, by putting a premium on good farming and encouraging the young to find careers for themselves in rural industries."

12. *Local institutions* of all kinds must have a powerful effect in evolving a good community sense. This is true in a superlative degree of the school, the church, the fair, and the rural library. These institutions will bring into the community the best thought of the world and will use it in the development of the people in the locality.

Such institutions must do an extension work. The church, from the nature of its organization, could readily extend itself beyond its regular and essential gospel work. The high-school will hold winter-courses and will take itself out to its constituency. The library ought to occupy its whole territory (page 92).

Similarly, village improvement societies should organize country and town together, extending tree-care, better roads, lawn improvement, and other good work throughout the entire community contributory to the city. Civic societies, fraternal orders, hospital associations, business organizations (page 119), women's clubs and federations, could do the same.

13. *The local rural press* ought to have a powerful influence in furthering community action. Many small rural newspapers are meeting their local needs, and are to be considered among the agents that make for an improved country life. In proportion as the support of the country newspaper is provided by political organizations, hack politicians, and patent medicine advertisements, will its power as a public organ remain small and undeveloped.

14. The influence of the *many kinds of extension teaching* is bound to be marked. Reading-courses, itinerant lectures, the organizing of boys' and girls' clubs, demonstration farms, the inspections of dairies, orchards, and other

farms, and of irrigation supplies, the organization of such educational societies as cow-testing societies, and the like, touch the very core of the rural problem. The influence of the traveling teacher is already beginning to be felt, and it will increase greatly in the immediate future. I mean by the traveling teacher the person who goes out from the agricultural college, the experiment station, the state or national department of agriculture, or other similar institutions, to impart agricultural information, and to set the people right toward their own problems.

15. The modern extension of *all kinds of communication* will unite the people, even though it does not result in making them move their residences. I have in mind good highways, telephones, rural free deliveries, and the like. The automobile is already beginning to have its effect in certain rural communities, but we have yet scarcely begun to develop the type of auto-vehicle which is destined, I think, to make a very great change in country affairs. The improvement of highways on a regular plan will itself tend to organize the rural districts.

We must add to all this a thoroughly developed system of parcels post, not only that the farmer may receive mail, but that he may also have greater facilities and freedom to transact his business with the world (page 118).

16. *Economic or business coöperation* must be extended. There is much coöperation of this kind among American farmers, more than most persons are aware. Some of it is very effective, but much of it is coöperative only in name. It takes the form of milk organizations, creameries, fruit associations, poultry societies, farmers' grain elevators, unions for buying and selling, and the like, some of which are of great extent.

A really coöperating association is one in which all members take active part in government and control, and share in their just proportions in the results. It is properly a society, rather than a company. Many so-called coöperative units are really stock companies, in which a few persons control, and the remainder become patrons; and others are mere shareholding organizations.

Business coöperation in agriculture is of three kinds: (1) coöperative production; (2) coöperative buying; (3) coöperative selling. The last two are extensively practiced in many regions. Coöperative production of animals and crops is practically unknown in the rural communities in the United States, and we are not to expect it to arise in those communities to any extent under the present organization of society. Colonies organized on a coöperative basis may practice it within their membership, but it is doubtful whether persons who are well equipped to be farmers will enter such organizations for this purpose so long as it is so easy to make a financial success at independent farming.

There is a fourth form that should be mentioned, although it is not coöperation in the real sense, but rather a form of combination. I refer to movements to control the production or output of commodities, as of wheat, cotton, tobacco, maize, and arbitrarily to fix the price. This cannot be permanently accomplished with any of the great staples, and even if it could be accomplished, in my opinion it would be an economic and social error.

Very much has been said about the necessity of business coöperation among farmers, and the importance of the subject can hardly be overstated; and yet it should be understood that economic coöperation is only one of many means that may be put in operation to propel country life. The essential thing is that country life be organized: if the organization is coöperative, the results — at least theoretically — should be the best; but in one place, the most needed coöperation may be social, in another place educational, in another religious, in another political, in another sanitary, in another economic in respect to buying and selling and making loans or providing insurance. When the chief deficiency in any region is economic, then it should be met by an organization that is primarily economic. Some of the effective coöperation in the West, so often cited, is really founded on the land-selling spirit of the community.

In some parts of the United States, the

financial status of the farmer is very low, but in general the economic condition is in advance of other conditions. The American farmer is prosperous,—not as prosperous as he ought to be, but so prosperous that he can conduct his own business without support or aid of his neighbors. Although he might gain financially by coöperation in any case, he nevertheless desires his complete freedom of action, even at the risk of some loss. The psychology of the American farmer is in the end the determining factor.

In other countries, this may not be so true, and particularly not when the farmers live under such a condition of peasanthood (or do not comprise a middle class) that no one of them in a community is able independently to buy his tools or his live-stock, or to secure sufficient funds to provide a small working capital, when both sales and purchases are very small, and when the entire community is practically subjugated by a political system. The big people are more likely to combine than to coöperate. Close coöperation naturally works best in a peasantry and under a paternal government; it becomes a means of bringing up the peasantry, of relieving them of oppression, and of giving them the rights that should be theirs as a part of their citizenship.

In Denmark, the coöperative movement has been one means of the salvation of the country, following the disastrous German war. The movement in some parts of the world is really a culture movement, having for a background the general good of society.

The American white farmer is not a peasant; he is not submerged in a hopeless political and economic slavery; he has his vote, his free school, his fee to hold property without let or hindrance, his full right to make the most of himself, his "rights" (pages 100 and 65). I think it will be possible for him to exercise these privileges and at the same time to share the benefits of coöperation; but coöperation is not necessary to win him these privileges. It is not the unit in his life, not the nucleus out of which all other agencies must evolve, or the leaven that will raise the lump: it is itself

K

one coördinating part in a program of evolution. We do not have the problem of peasant proprietorship. For the most part, the American farmer has already won his economic independence, if not his just rewards.

We should not be impatient if our farmers do not organize themselves coöperatively as rapidly as we think they ought to organize.

Economic personal coöperation may be expected to thrive best in a community of small farmers. It is a question whether we shall develop the strongest leaders in a condition of more or less uniform small farms. There is much to be said in favor of rather large farming (say 500 to 1000 acres), for a business of this proportion demands a strong man. This does not mean landlordism, which is a part of a political and hereditary system, but merely large and competent business organization. Such farmers, if they are so minded, can accomplish great things for their fellows.

I am looking for some of the best results in coöperation to come from the establishment of field-laboratories and demonstration farms, to which the farmers of the locality contribute their personal funds in the expectation of an educational result. The best results to country life cannot possibly come by the government continuing to take everything to the farmer free of cost and without the asking. Disadvantaged or undeveloped regions must be aided freely, but as rapidly as any localities or industries get on their feet, they should meet the state part way, and should assume their natural share of the expense and responsibility. This form of coöperation is already well under way; and I suspect that in many localities that have been dead to all forms of coöperative effort, this idea will afford the starting-point for a new community life.

From this form of education-coöperation, it would be but a step to a neighborhood effort to introduce new crops and high-class bulls, to undertake drainage enterprises and reforestation; and to unite on business matters.

It is possible for a national organization movement to come out of the existing agricultural institutions in the United States.

We may picture to ourselves a perfectly coöperating rural society that will have all the means of its salvation within itself. Even if we accept this picture, we cannot say that the structure will rise out of one seed or starting-point, or that one phase of coöperation is of necessity primary and another final. Our theoretical structure will arise from several or many beginnings; it will be a complex of numberless units; whatever range of coöperation is found, by investigation, to be now most needed in any community, must be the one with which we are to set that community going.

17. In the end everything depends on *personal gumption and guidance*. It is not strange that we have lacked the kind of guidance that brings country people together, because we have not had the kind of education that produces it; and, in fact, this kind of guidance has not been so necessary in the past as it is now. A new motive in education is gradually beginning to shape itself. This must produce a new kind of outlook on country questions, and it will bring out a good many men and women

who will be guides in the country as their fellows will be guides in the city. They will be captains because they will perform the common work of farming regions in an uncommon way.

I think we little realize to-day what the effect will be in twenty-five years of the young men and women that the colleges of agriculture in these days are sending into the country districts.

Community interest is of the spirit.

In conclusion, let us remember that everything that develops the common commercial, intellectual, recreative, and spiritual interests of the rural people, ties them together socially. Residing near together is only one of the means of developing a community life, and it is not now the most important one. Persons who reside close together may still be torn asunder by divergent interests and a simple lack of any tie that binds; this is notably true in many country villages.

Community of purpose and spirit is much more important than community of houses. Community pride is a good product; it produces a common mind.

A POINT OF VIEW ON THE LABOR PROBLEM

It is a general complaint in the United States that there is scarcity of good labor. I have found the same complaint in parts of Europe, and Europeans lay much of the blame of it on America because their working classes migrate so much to this country; and they seem to think we must now be well supplied with labor. Labor scarcity is felt in the cities and trades, in country districts, in mines, and on the sea. It seems to be serious in regions in which there is much unemployed population. It is a real problem in the Southern states.

While farmers seem now to complain most of the labor shortage, the difficulty is not peculiarly rural. Good farmers feel it least; they have mastered this problem along with other problems. As a matter of fact, it is doubtful whether there is a real labor shortage as measured by previous periods; but it is very difficult to secure good labor on the previous terms and conditions.

Reasons for the labor question.

The supposed short labor supply is not a temporary condition. It is one of the results of the readjustment and movement of society. A few of the immediate causes may be stated, to illustrate the nature of the situation.

(1) In a large way, the labor problem is the result of the passing out of the people from slavery and serfdom,—the rise of the working classes out of subjugation. Peoples tend always to rise out of the laboring-man phase. We would not have it otherwise if we desire social democracy.

(2) It is due in part to the great amount and variety of constructive work that is now being done in the world, with the consequent urgent call for human hands. The engineering and building trades have extended enormously. We are doing kinds of work that we had not dreamed of a half-hundred years ago.

(3) In some places the labor difficulty is due to the working-men being drawn off to other places, through the perfecting of industrial organization. The organization of labor means companionship and social attraction. Labor was formerly solitary; it is now becoming gregarious.

(4) In general, men and women go where things are "doing." Things have not been doing on the farms. There has been a gradual passing out from backward or stationary occupations into the moving occupations. Labor has felt this movement along with the rest. It has been natural and inevitable that farms should have lost their labor. Cities and great industrialism could not develop without them; and they have made the stronger bid.

(5) In farming regions, the outward movement of labor has been specially facilitated by lack of organization there, by the introduction of farm machinery, by the moving up of tenants into the class of renters and owners, by lack of continuous employment, by relatively low pay, by absence of congenial association as

compared with the town. Much of the hired farm labor is the sons of farmers and of others, who "work out" only until they can purchase a farm. Some of it is derived from the class of owners who drift downward to tenants, to laboring men, and sometimes to shifters. We are now securing more or less foreign-born labor on the farms. Much of this is merely seasonal; and when it is not seasonal, the immigrant desires to become a farm owner himself. If the labor is seasonal, the man may return to his native home or to the city, and in either case he is likely to be lost to the open country.

The remedies.

There is really no "solution" for the labor difficulty. The problem is inherent in the economic and social situation. It may be relieved here and there by the introduction of immigrants or by transportation of laborers at certain times from the city; but the only real relief lies in the general working out of the whole economic situation. The situation will gradually correct itself; but the readjustment will come much more quickly if we understand the conditions.

As new interest arises in the open country and as additional values accrue, persons will remain in the country or will return to it; and the labor will remain or return with the rest. As the open country fills up, we probably shall develop a farm artisan class, comprised of persons who will be skilled workmen in certain lines of farming as other persons are skilled workmen in manufactures and the trades. These persons will have class pride. We now have practically no farm artisans, but solitary and more or less migratory working-men who possess no high-class manual skill. Farm labor must be able to earn as much as other labor of equal grade, and it must develop as much skill as other labor, if it is to hold its own. This means, of course, that the farming scheme may need to be reorganized (pages 86 to 90).

Specifically, the farm must provide more continuous employment if it is to hold good labor. The farmer replies that he does not have employment for the whole year; to which the answer is that the business should be so reorganized as to make it a twelve months' enterprise. The introduction of crafts and local manufactures will aid to some extent, but it cannot take care of the situation (page 115). In some way the farm laborer must be reached educationally, either by winter schools, night schools, or other means. Every farm should itself be a school to train more than one laborer. The larger part of the farm labor must be country born. With the reorganization of country life and its increased earning power, we ought to see an increase in the size of country families.

Public or social bearings.

It is doubtful if city industrialism is developing the best type of working-men, considered from the point of view of society (page 59). I am glad of all organizations of men and women, whether working-men or not. But it seems to me that the emphasis in some of the organizations has been wrongly placed. It has too often been placed on rights rather than on duties. No person and no people ever developed by mere insistence on their rights. It is responsibility that develops them. The working-man owes responsibility to his employer and to society; and so long as the present organization of society continues he cannot be an effective member of society unless he has the interest of his employer constantly in mind.

The real country working-men must constitute a group quite by themselves. They cannot be organized on the basis on which some other folk are organized. There can be no rigid short-hour system on a farm. The farm laborer cannot drop his reins or leave his pitchfork in the air when the whistle blows. He must remain until his piece of work is completed; this is the natural responsibility of a farm laborer, and it is in meeting this responsibility that he is able to rise to the upper grade and to develop his usefulness as a citizen.

It is a large question whether we are to have a distinct working-class in the country as distinguished from the land-owning farmer. The old order is one of perfect democracy, in which the laboring-man is a part of the farmer's family. It is not to be expected that this condition can continue in its old form, but the probability is that there will always be a different relation between working-man and employer in the country from that which obtains in the city. The relation will be more direct and personal. The employer will always feel his sense of obligation and responsibility to the man whom he employs and to the man's family. Persons do not starve to death in the open country.

Some persons think that the farming of the future is still to be performed on the family-plan, by which all members of the family perform the labor, and whatever incidental help is employed will become for the time a part of the family. This will probably continue to be the rule. But we must face the fact, however, that a necessary result of the organization of country life and the specialization of its industries, that is now so much urged, will be the production of a laboring class by itself.

Supervision in farm labor.

It is doubtful whether we shall extend the industrial organization of labor to the open country, and yet there should be some way of administering farm labor. The growth of the tendency to coördinate farming industries, in order to overcome the disastrous effects of much of the competitive farming, will allow for supervision of labor, however, and will make for efficiency. The standardizing of agricultural practice will also do much to produce the community mind that is so much desired (p. 97). On this line, Dean H. E. Cook, who has given much thought to labor questions, writes me as follows:

"The production of iron, paper, and manufactured products generally has been standardized, and the cost laid down in the market is well known, and therefore placed squarely on a cash basis. Directly the opposite is the case in the manufacture of farm crops, and so we find the family to be the farm crop-producers. The wife and the children are a part of the working force of the farm, which is not found in any other industry. In fact, our laws are very rigid in preventing the employment of women and children in nearly every class of work, except on the farm. We find no provision by statute or moral sentiment which says that the farmer must not employ his eight- or ten-year-old boy, as is very often the case, in most laborious tasks. This state of affairs is not the desire of the farmer, but has become a necessity because of the very low prices for his products, occasioned by the intense competition of the rapidly extending area. Our government has taken every means within its grasp to populate these large areas of cheap rich land. Of course it meant wealth to the nation, but it meant poverty to those who had established homes and investments in the older sections.

"Our methods, unlike other manufacturers and producers, are not standardized. That is, we find in every community persons having each his own conception of soil-handling, crop-growing, and marketing. In a single locality can be found an endless variety of corn, as an illustration. Especially is this true in the East.

Surely corn growing fourteen feet high and corn growing six feet high are not calculated to bring the same results. The farmers themselves are unlike. I suppose we are distantly removed from the time when we shall have a uniform type of men and women bred for the farm. It seems to me that methods which would unify or standardize our practices and prices — within certain limitations, to be sure — would tend to unify the tendencies and the type of the people.

"In our present state of undevelopment or adjustment, I do not think it is possible profitably to pursue the production of crops with employed labor, such as we find in our manufacturing establishments; and it may be debatable whether that plan would be an improvement, so far as the social life is concerned, over the present family-plan, although I firmly believe that the time is approaching when the profits of the business will warrant a cash payment for everything done on the farm. As a connecting link between the family-plan and the future cash-plan, it seems to me we ought to take on in each neighborhood the same methods of supervision that are now employed in the factories. One man of skill and adaptability supervises the work of many. In agriculture we have but one illustration of this principle, namely, our butter and cheese factories, where one man has in charge the manufacturing of the milk of many. I think we could profitably use a similar agency in trucking, soil-handling, crop-growing, animal-feeding, and general farm-management. Furthermore, we are more in need, as the writer sees it, of this standardizing or coöperation in farm-management, than we are in the manufacture of milk products. This plan would use the family as a unit of labor on the farm, with the attendant light risk, or no risk at all; and in case of failure of crops of having to pay cash for the labor.

"The cow-test association is a part of this general plan of local supervision. I can foresee how there may come out of this cow-test movement, a growth which will mean just what I have tried to outline. The man who does nothing now but the testing of the milk from each cow may develop into an expert who will give advice on soils, crops, cow-feeding, and other things (page 123).

"When the communities around certain natural centers, as the cheese factories or creameries in dairy sections, perhaps a small hamlet in trucking sections, have become thoroughly organized or, more properly speaking, standardized, we shall find it comparatively easy to bring a number of these local units together, because the individuals who form a part of the movement have learned the true principles underlying coöperation. Until these local units are worked out, in my opinion we shall never be able to form any great coöperative movement which will not break of its own weight, because of a lack of annealing processes."

What is the farmer to do?

"How may I secure labor?" is probably the most persistent question now asked by farmers; but it is a question that cannot be answered, any more than one may tell another what crops he shall grow, what markets he shall find, or what manner of house he shall build. This is one of the great problems of farming, as it is of engineering, of the building trades, and of factories. Each farmer must work it out for himself, as he works out the problem of fertility and machinery. He must work far ahead, and consider it as a part of all his plans.

In many or most cases, it resolves itself into a question of personality, — of making a place that is worth while to a good man and then of the farmer interesting himself in the man. One can now hardly expect to secure labor on demand for brief periods, for the scheme of things is more and more in the direction of continuous employment; and the old range of prices cannot hold. If the farmer's scale of business is small and operates only for a part of a year, he cannot expect to secure the best and most reliable help.

The farmer will find increasing aid from public labor-distributing bureaus, for these agencies must extend with the extension of population and the complexity of industry. In time, the state and nation will provide competent machinery for placing working-men where they can best serve themselves and society, thus relieving both employer and employed from much waste of effort. As farm labor is not a separate difficulty, the problem will tend to better and better solution along with the rest. If the distributing agencies are not now wholly satisfactory, the farmer must recognize that they are only beginning, and that he should coöperate with them. The problem of utilizing the immigrant, for example, is one of distribution; but distribution is really not accomplished merely by sending a certain number of immigrants to a certain number of places, — immigrant and employer must find the situation to be mutually satisfactory.

Any effort which assumes that labor must necessarily come to the old-type farm, is only temporary. The farm must readjust itself to meet the labor problem. In the meantime, through the labor bureaus, by looking long ahead, by organizing a labor club in the community, by some person acting as a labor agent and supplying farmers as they need, by trying to make a year-round activity in the neighborhood, the situation may be met more or less.

THE MIDDLEMAN QUESTION

To make farming profitable is no longer a question merely of raising more produce. We have passed that point. We now have knowledge and experience enough to enable us greatly to increase our yields, if only we put the knowledge into practice.

Farmer does not get his share.

But the farmer, speaking broadly, does not get his share of the proceeds of his labor, notwithstanding the increase in the price of farm products. A few farmers here and there, producing a superior article and favored by location or otherwise, can be quite independent of marketing systems; but the larger number of farmers never can be so situated, and they must grow the staples, and they are now at the mercy of many intermediaries. The farmer's risks, to say nothing of his investment and his labor, are not sufficiently taken into account in our scheme of business,—risks of bad years, storm, frost, flood, disease to stock and crop, and many things over which he has practically no control.

A merchant in a small city may want as much as twenty per cent commission to sell produce, and then retain the privilege of returning to the grower all the product that spoils on his hands or that he is unable to sell; he invests little capital, takes no risk, and makes more than the man who buys his land, prepares the crop months in advance, and assumes every risk from seed-time to dinner-table. I am citing this case not to say that it is a subject for public control nor even to assert that the merchant's commission is intrinsically too great, but only to illustrate the disadvantage in which the farmer often finds himself; and the farmer may even have no escape from this disadvantage, for all the merchants within his market region may agree to sell his produce only on such terms, and he may be obliged to accept these terms or not to sell his wares.

The manufacturer knows the cost of his products and charges his price. The farmer usually does not know the cost, and in general he makes no selling price; the prices of his staple produce are made for him.

That the producer does not secure his proportionate share of the selling price in many products is a matter of the commonest knowledge, and much study has been made of the question. If the question is put in another way, the consumer pays too great a margin, in great numbers of cases, over the cost of production. The following press item, coming to my hand as I write, is an example (given for what it is worth), although not extreme: "The government of New York, and not the government in Washington, is where the people of this city must look, if they expect to see reduction in living expenses. A bushel of beans, for which the producer in Florida receives $2.25, with the transportation 50 cents for the 800-mile haul, should not cost the New York consumer $6.40 a bushel. The producer receives 35 per cent of the final price, the transporter 8 per cent, and the dealers 57 per cent. This is not a fair division. The problem is not one of trusts, tariffs, and other Washington matters, but simply one of providing straight and cheap ways open from all gardens and farms to kitchens and tables."

The poorer the country or the less forehanded the people, the harder is the pinch of the usurer and the trader, and all the machinery of trade is likely to be manipulated against the defenseless man who stands stolidly between the handles of the plow.

Of course, such conditions do not obtain with all products. In some of the great staples, as wheat, the cost of transportation and commissions is often reduced by competition and scientific handling to probably its lowest terms. But that there are abuses and extortions, and remediable conditions, in the middleman system — by which I mean collectively all traders between producer and consumer — no one will attempt to deny. The farmer cannot rise to his proper place until the stones are taken off his back.

The abuses must be checked and discriminations removed, whether in the middleman trade itself, rates of express companies and other carriers, or stock-market gambling. The middleman system has had a free field to play in, the wealth of the country to handle; it has exercised its license, and in too many cases it has become parasitic, either protected by law and custom or unreachable by law or custom. It is a shame that our economic machinery is not capable of handling the situation.

Relation of the question to cost-of-living.

It is customary just now to attribute the high cost of living to lessened production due to a supposed decline of agriculture, and to advise, therefore, that more persons engage in farming for the purpose of increasing the product. This position is met by an editorial of the New York Tribune, which holds that intermediary trading combinations are responsible:

"It is true that the raising of cattle for the market has almost ceased in the East and that agriculture generally has not kept pace with the demand for food products. Yet it is hard to believe that agriculture in any part of the Union would steadily decline in the face of an enormous appreciation of the cost to the consumer of all farm products, were there not some powerful disturbing factor operating to deny the farmer the benefits of that appreciation. If the Eastern farmer could have reaped a legitimate share of the increase in the price of farm produce which has taken place in the last twenty years, he would certainly be in position to command all the labor he needs and to develop resources now neglected because it does not pay to develop them. Under normal conditions economic law would certainly drive labor and energy into a field of production in which there had been the greatest relative expansion in the selling price of products.

"Yet economic law has not operated to stimulate agriculture, because the returns from steadily mounting prices have not really reached the producer. Thirty years ago the fattening of steers for the local markets was common in the

East. But when the vast Western ranges were opened, and the great packing houses were established, the cheapness of range beef, refrigerated and delivered in Eastern cities, was used as a weapon to kill off the cattle industry of the East. When the Eastern cattleman was driven out of business, the price of beef rose, but virtually all the increase has gone to the packing combinations, which fix their own price to the Western range man and their own price to the consumer and artificially control the supply so as to discourage increased production in the West and to prevent a revival of production in the East. The country is growing in population at the rate of twenty to twenty-five per cent each decade. But Secretary Wilson has shown that the supply of food animals is not being maintained in proportion to population. In the last decade cattle have remained about stationary in numbers, swine are actually decreasing, and, while more sheep are available, the supply has diminished relatively to population.

"It can hardly be contended that with steadily diminishing supplies and steadily increasing prices the law of supply and demand would not work out a new balance, stimulating production through easy profits, were there no artificial interception of the producer's normal share of the advance in price. Were there a free market for the Eastern raiser of stock, milk, and food products generally, with the middleman's commissions properly restricted, Eastern farming would probably be profitable enough to hold its own against manufacturing and to compete successfully with the manufacturer for labor."

The farmer's part.

Of course, it is necessary to teach every farmer how to grow more crops, for this is his business, and it also enlarges his personal ambition and extends his power and responsibility; but merely to grow the crops will not avail, — this is only the beginning of the problem: the products must be distributed and marketed in such a way that the one who expends the effort to produce them shall receive enough of the return to identify him with the effort. Thereafter, social and moral results will follow.

The middleman's part.

I recognize the service of the middleman to society. I know that the distributor and trader are producers of wealth as well as those who raise the raw materials; but this is no justification for abuses. I know that there are hosts of perfectly honest and dependable middlemen. We do not yet know whether the existing system of intermediary distributors and sellers is necessary to future society, but we do not see any other practicable way at present. In special cases, the farmer may reach his own customer; but this condition, as I have suggested, is so small in proportion to the whole number of farmers as not greatly to affect the general situation. We do not yet see any way whereby all farmers can be so organized as to enable them to control all their own marketing. Therefore, we must recognize middleman-practice as legitimate.

A system of economic waste.

But even though we yet see no way of general escape from the system, we ought to provide some means of regulating its operation. The present method of placing agricultural produce in the hands of the consumer is for the most part indirect and wasteful. Probably in the majority of cases of dissatisfaction, the person whom we call the middleman does not receive any exorbitant profit, but the cost of the commodities is piled up by a long and circuitous system of intermediate tolls and commissions.

Coöperation of farmers will not solve it.

It is commonly advised that farmers "unite" or "organize" to correct middleman and transportation abuses, but these troubles cannot be solved by any combination of farmers, because this is not an agricultural question. It is as much a problem for consumers as for producers. It is a part of the civilization of our day, completely woven into the fabric of our economic system. The farmer may feel its hardship first because he must bear it, while the consumer, to meet higher prices, demands more pay of his employer or takes another stitch out of somebody else. But it is essentially a problem for all society to solve, not for farmers alone, particularly when it operates on a continental basis. This also indicates the futility of the arbitrary control of prices of the great staples by combinations of farmers (page 126).

Of course, temporary or local relief may be secured by organizations of producers here and there, or of consumers here and there (probably consumers can attack the problem more effectively than producers), and by the establishment of public markets; but no organization can permanently handle the question unless the organization is all the people.

The present agitations against middleman practices and stock-market gambling ought to compel Congress to pass laws to correct the evils that are correctable by law, and the organizations then should keep such touch on the situation that the laws will be enforced.

It has been suggested that the superabundant middlemen go into farming; but no one can compel them to go to farming, and they might not be successful farmers if they should attack the business, and the farming country might not need them or profit by them,— for it is not demonstrated that we need more farmers, although it is apparent that we need better farmers.

It is the business of government.

It is the business of any government to protect its people. Governments have protected their countries from invasion and war, but the greatest office of government in modern times is to develop its own people and the internal resources of its realm. We are beginning to protect the people from the over-lording of railroads, from unfair combinations in trade, and from the tyranny of organized politicians. It is just as much the business of government to protect its people from dishonest and tyrannous middlemen lying beyond the practical reach of individuals. The situation has arisen because of lack of control; there is no conspiracy against the farmer.

It is said that competition will in the end correct the middleman evil, but competition does not correct it; and competition alone, under the present structure of society, will not correct it in most cases because "agreements" between traders restrict or remove competition: the situation does not have within itself the remedies for its own ills.

When we finally eliminate combinations in restraint of trade, the middleman abuses may be in the process of passing out. It is to check dishonesty on the one hand and to allow real competition on the other that I am now making suggestions.

Must be a continuing process of control.

I have no suggestion to make as to the nature of the laws themselves. There are many diverse situations to be met; and I intentionally do not make my remarks specific. Of course, any law that really attempts to reach the case must recognize the middleman as exercising a public or semi-public function, and that, as such, he is amenable to control, even beyond the point of mere personal honesty. The licensing of middlemen (a practice that might be carried much further, and which is a first step in reform) recognizes this status; and if it is competent for government to license a middleman, it is also competent for it to exercise some oversight over him. It is not necessary that government declare an agency a monopoly in order to regulate it. Commercial situations that unmistakably involve service to the public are proper for governmental control in greater or lesser degree. The supervision of weights and measures is a good beginning in the regulation of middleman trading.

But the enactment of laws, even of good laws, is only another step in the solution. A law does not operate itself, and the common man cannot resort to courts of law to secure justice in such cases as these. There must be a *continuing process* of government with which to work out the reform and to adjust each case on its merits. Whatever the merits of the laws, their success lies in the continuing application of them to specific cases by persons whose business it is to discern the facts rather than to prove a case.

There are three steps in the control of the middleman: (1) an aroused public conscience on the question; (2) good fundamental laws for interstate phases and similar state laws for local phases; (3) good commissions or other agencies or bodies to which any producer or consumer or middleman may take his case, and which may exercise regulatory functions. The interstate commerce commission has jurisdiction over so much of the problem as relates to the service and rates of common carriers; no doubt, its powers could be extended to other interstate phases. Perhaps departments of agriculture, in states in which public service commissions have not been established, could be given sufficient scope to handle some of the questions.

Of course, some of the middlemen and associated traders will contend that all this interferes with business and with private rights, but no man has a private right to oppress or defraud another or to deprive him of his proper rewards; and we must correct a faulty economic system.

There is little danger that the legitimate business of any honest middleman will be interfered with.

I know that commissions and similar bodies have not always been wholly successful. This is because we have not yet had experience enough, have not consciously trained our people for this kind of work, and have not been able to make water-tight laws. Neither do older systems now prove to be adequate. New economic conditions must bring new methods of regulation and control.

I have no desire that society (or government) engage in the middleman business or that it take over private enterprise; but no government can expect to throw back on the producer the responsibility of controlling the middleman. I look for the present agitation to awaken government to the necessity of doing what it is plainly its duty to do. In future, a government that will not protect its people in those cases in which the people, acting to the best of their individual and coöperating capacity, cannot protect themselves, will be known as either a bad government or an undeveloped government.

COUNTY AND LOCAL FAIRS

Much is said about the necessity of redirecting rural institutions. The fairs are mentioned among the rest. I shall now indicate an experiment that might be tried with existing county and local fairs, not only as a suggestion for the fairs themselves, but as an illustration of how completely it is possible to reconstruct an institution that is long established in conventional methods.

I do not think a fair that carries only one or two weeks' interest during the year is justifiable; but of this aspect of the question I am not now speaking.

Nature of the fair.

The county fair has not changed its general basis of operation in recent years, and yet the basis of country life is changing rapidly. Many fairs are doing excellent work and are worth to the people all that they cost in effort and money; but the whole plan of the county fair is insufficient for the epoch that we are now entering. I should not discontinue the local fairs: I should make them over.

The fairs have been invaded by gambling, and numberless catch-penny and amusement and entertainment features, many of which are very questionable, until they often become great country medleys of acrobats and trained bears and high-divers and gew-gaws and balloon ascensions and side-shows and professional traveling exhibitors and advertising devices for all kinds of goods. The receipts are often measured by the number of cheap vaudeville and other "attractions" that the fair is able to secure. And as these things have increased, the local agricultural interest has tended to drop out. In some cases the state makes appropriations to local fairs; it is a question whether the state should be in the showman business.

I should like to see one experiment tried somewhere by some one, designed to project a bold enterprise on a new foundation. It would first be necessary to eliminate some of the present features, and then to add a constructive program.

Features to be eliminated.

I should eliminate all gate receipts; all horse trots; all concessions and all shows; all display of ordinary store merchandise; all sales of articles and commodities; and all money premiums.

Constructive program.

Having taken out the obstructions, unnecessaries, and excrescences, I should enter on a constructive program. I should then begin to make a fair. I assume that the fact of a person living in a community, places on him responsibilities for the welfare of that community. We should make the county fair one of the organized means of developing this welfare. Therefore, I should assume that every citizen in the county, by virtue of his citizenship, is a member of the county fair and owes to it an allegiance.

It would then devolve on the persons who

are organizing and operating the work, representing the fair association, to develop in him his sense of allegiance and coöperation. I should not discourage any citizen of the county from coöperating in the enterprise, or allow him to escape his natural responsibilities, because he felt himself unable or unwilling to pay an admission fee, any more than I should eliminate any person because of religion, politics, color, or sex.

The financial support.

Of course, it requires money to run a fair. I should like to see the money raised by voluntary contribution in a new way. I should have it said to every resident in the county that he and his family may come uninterruptedly to the fair without money and without price; but I should also say to him that money is needed, and that all those persons who wish to give a certain sum would be provided with a badge or receipt. I suspect that more money could be more easily raised in this way than by means of gate receipts.

I should have this money collected in advance by means of an organized effort through all the schools and societies in the county, setting every one of them at work on a definite plan.

Of course, the state or other agency could contribute its quota of funds as theretofore.

An educational basis.

In other words, I should like to see, in this single experiment, a complete transfer from the commercial and "amusement" phase to the educational and recreation phase. I should like to see the county fair made the real meeting place for the country folk. I should make a special effort to get the children. The best part of the fair would be the folks, and not the machines or the cattle, although these also would be very important. I should make the fair one great picnic and gathering-place and field-day, and bring together the very best elements that are concerned in the development of country life.

I should work through every organized

enterprise in the county, as commercial clubs, creameries, coöperative associations, religious bodies, fraternal organizations, insurance societies, schools, and whatever other organized units already may exist.

It is often said that our fairs have developed from the market-places of previous times, and are historically commercial. We know, of course, that fairs have been market-places, and that some of them are so to this day in other countries. I doubt very much, however, whether the history is correct that develops the American agricultural fair from the market-place fairs of other countries. From the time when Elkanah Watson exhibited his merino sheep in the public square of Pittsfield, Massachusetts, in 1807, in order that he might induce other persons to grow sheep as good as his, and when the state of New York started its educational program in 1819, the essence of the American idea has been that a fair is an educational and not a trading enterprise. But whatever the history, the agricultural fair maintained by public money owes its obligation to the people and not to commercial interests.

Ask every person to prove up.

I should have every person bring and exhibit what he considers to be his best contribution to the development of a good country life.

One man would exhibit his bushel of potatoes; another his Holstein bull; another his pumpkin or his plate of apples; another a picture and plans of his modern barn; another his driving team; another his flock of sheep or his herd of swine; another his pen of poultry; another his plan for a new house or a sanitary kitchen, or for the installation of water-supplies, or for the building of a farm bridge, or the improved hanging of a barn door, or for a better kind of fence, or for a new kink in a farm harness, or the exhibition of tools best fitted for clay land or sandy land, and so on and on.

The woman would also show what she is contributing to better conditions, — her best handiwork in fabrics, her best skill in cooking, her best plans in housekeeping, her best ideas for church work or for club work.

The children would show their pets, what they had grown in the garden, what they had made in the house or the barn, what they had done in the school, what they had found in the woods.

I should assume that every person living on the land in the country has some one thing that he is sure is a contribution to better farming, or to better welfare; and he should be encouraged to exhibit it and to explain it, whether it is a new way to hang a hoe, or a herd of pure-bred cattle, or a plan for farmers' institutes. I should challenge every man to show in what respect he has any right to claim recognition over his fellows, or to be a part of his community.

I should ask the newspapers and the agricultural press to show up their work; also the manufacturers of agricultural implements and of country-life articles of all kinds.

I should also ask the organizations to prove up. What is the creamery contributing to a better country life? What the school? The church? The grange? The coöperative exchange? The farmers' club? The reading club? The woman's society? The literary circle? The library? The commercial clubs? The hunting or sportsman's clubs?

Sports, contests, and pageants.

I should give much attention to the organization of good games and sports, and I should have these coöperative between schools, or other organizations, such organizations having prepared for them consecutively during the preceding year. I should introduce good contests of all kinds. I should fill the fair with good fun and frolic.

I should want to see some good pageants and dramatic efforts founded on the industries, history, or traditions of the region or at least of the United States. It would not be impossible to find simple literature for such exercises even now, for a good deal has been written. By song, music, speaking, acting, and various other ways, it would not be difficult to get all the children in the schools of the county at work. In the old days of the school "exhibition," something of this spirit prevailed. It was manifest in the old "spelling bees" and also in the "lyceum." We have lost our rural cohesion because we have been attracted by the town and the city, and we have allowed the town and the city to do our work. I think it would not be difficult to organize a pageant, or something of the kind, at a county fair, that would make the ordinary vaudeville or sideshow or gim-crack look cheap and ridiculous and not worth one's while.

Premiums.

If we organize our fair on a recreation and educational basis, then we can take out all commercial phases, as the paying of money premiums. An award of merit, if it is nothing more than a certificate or a memento, would then be worth more than a hundred dollars in money. So far as possible, I should substitute coöperation and emulation for competition, particularly for competition for money.

It is probable that the fair would have to assume the expense of certain of the exhibits.

It is time to begin.

This kind of fair is not only perfectly possible, but it is feasible in many places, if only some one or two or three persons possessed of good common sense and of leadership would take hold of the thing energetically. One must cut himself loose from preconceived notions and probably from the regular fair associations. He must have imagination, and be prepared to meet discouragements. He need not take the attitude that present methods are necessarily all bad; he is merely concerned in developing a new thing.

Because I should not have horse races in my fair, I do not wish at all to be understood as saying that horse races are to be prohibited. Let the present race courses in the fair grounds be used for horse races, if the people want them. We have June races now, and they could be held at other times of the year when persons who are interested desire to have them. My point is that they are not an essential part of a county agricultural fair.

They rest on a money basis, and do not represent the people. Neither do I say that all traveling shows and concessions are bad; but most of them are out of place in a county fair and contrary to its spirit.

If the horse races were organized for the purpose of developing the horses of the county, then I should admit them; but I should give them only their proportionate place along with other means of developing horse-stock, — as of work horses, farm horses, draft horses, driving horses.

The fair ground.

An enterprise of the kind that I project need not necessarily be held on a fair-ground of the present type, although that might be the best place for it. If there is a good institution in the county that has grounds, and especially that has an agricultural equipment worthy of observation, I should think that the best results would be secured by holding the fair at that place. This kind of a fair would not need to be inclosed within a Chinese wall. Of course, there would have to be buildings and booths and stables in which exhibitions could be made.

In every fair there should nowadays be an assembly hall in which lectures, exhibitions, simple dramas, worth-while applicable moving pictures, and other entertainment features can be given.

My plea.

My plea, therefore, is that some one somewhere make one experiment with a county fair designed to bring all the people together on a wholly new idea. The present basis is wrong for this twentieth century. The old needs are passing; new needs are coming in. I should have the fair represent the real substantial progress of rural civilization, and I should also have it help to make that progress. It should be a power in its community, not a phenomenon that passes as a matter of course, like the phases of the moon.

I do not expect all this to materialize in a day; but I want to set a new picture into my readers' minds.

THE COUNTRY-LIFE PHASE OF CONSERVATION

The conservation movement is the expression of the idea that the materials and agencies that are part of the furniture of the planet are to be utilized by each generation carefully, and with real regard to the welfare of those who are to follow us. The country-life movement is the expression of the idea that the policies, efforts, and material well-being of the open country must be highly sustained, as a fundamental essential of a good civilization; and it recognizes the fact that rural society has made relatively less progress in the past century than has urban society. Both movements are immediately economic, but in ultimate results they are social and moral. They rest on the assumption that the welfare of the individual man and woman is to be conserved and developed, and is the ultimate concern of governments; both, therefore, are phases in a process of social evolution.

These are the twin policies of the Roosevelt administration, an economic and social movement for which that administration will be first remembered after the incidents and personalities of the time have lost their significance.

Not only the welfare but the existence of the race depends on utilizing the products and forces of the planet wisely, and also on securing greater quantity and variety of new products. These are finally the most fundamental movements that government has yet attempted to attack; for when the products of the earth shall begin to disappear or the arm of the husbandman to lose its skill, there is an end to the office of government. At the bottom, therefore, the conservation and country-life movements rest on the same premise; but in their operation and in the problems that are before them they are so distinct that they should not be confounded or united. These complementary phases may best work themselves out by separate organization and machinery, although articulating at every point; and this would be true if for no other reason than that a different class of persons, and a different method of procedure, attach to each movement. The conservation movement finds it necessary, as a starting-point, to attack intrenched property interests, and it therefore discovers itself in politics, inasmuch as these interests have become intrenched through legislation. The country-life movement lacks these personal and political aspects, and proceeds rather on a broad policy of definite education and of redirection of imagination.

These subjects have a history.

Neither conservation nor country life is new except in name and as the subject of an organized movement. The end of the original resources has been foreseen from time out of mind, and prophetic books have been written on the subject. The need of a quickened country life has been recognized from the time that cities began to dominate civilization; and the outlook of the high-minded countryman has been depicted from the days of the classical writings until now. On the side of mineral and similar resources, the geologists amongst us have made definite efforts for conservation; and on the side of soil fertility the agricultural chemists and the teachers of agriculture have for a hundred years maintained a perpetual campaign of conservation. So long and persistently have those persons in the agricultural and some other institutions heard these questions emphasized, that the startling assertions of the present day as to the failure of our resources and the coördinate importance of rural affairs with city affairs have not struck me with any force of novelty.

But there comes a time when the warnings begin to collect themselves, and to crystallize about definite points; and my purpose in suggesting this history is to emphasize the importance of the two formative movements now before us by showing that the roots run deep back into human experience. It is no ephemeral or transitory subject that we are now to discuss.

They are not party-politics subjects.

I have said that these are economic and social problems and policies. I wish to enlarge this view. They are concerned with saving, utilizing, and augmenting, and only secondarily with administration. We must first ascertain the facts as to our resources, and from this groundwork impress the subject on the people. The subject must be approached by scientific methods. The "political" phase, although probably necessary, is only temporary, till we remove impedimenta and clear the way.

It would be unfortunate if such movement became the exclusive program of a political party, for then the question would become partisan and probably be removed from calm or judicial consideration, and the opposition would equally become the program of a party. Every last citizen should be naturally interested in the careful utilization of our native materials and wealth, and it is due him that the details of the question be left open for unbiased discussion rather than to be made the arbitrary program, either one way or another, of a political organization. Conservation is in the end a plain problem involving economic, educational, and social situations, rather than a political issue.

The country-life movement is equally a scientific problem, in the sense that it must be approached in the scientific spirit. It will be inexcusable in this day if we do not go at the subject with only the desire to discover the facts and to arrive at a rational solution, by non-political methods.

The soil is the greatest of all resources.

The resources that sustain the race are of two kinds, — those that lie beyond the power of man to reproduce or increase, and those that may be augmented by propagation and by care. The former are the mines of minerals, metals, and coal, the water, the air, the sunshine; the latter are the living resources, in crop and livestock.

Intermediate between the two classes stands the soil, on which all living resources depend. While the soil is part of the mineral and earthy resources of the planet, it nevertheless can be increased in its producing power. Even after all minerals and metals and coal are depleted, the race may sustain itself in comfort and progress so long as the soil is productive, provided, of course, that water and air and sunshine are still left to us. The greatest of all resources that man can make or mar is the soil. Beyond all the mines of coal and all the precious ores, this is the heritage that must be most carefully saved; and this, in particular, is the country-life phase of the conservation movement.

To my mind, the conservation movement has not sufficiently estimated or emphasized this problem. It has laid stress, I know, on the enormous loss by soil erosion and has said something of inadequate agricultural practice, but the main question is yet practically untouched by the movement,— the plain problem of handling the soil by all the millions who, by skill or blundering or theft, produce crops and animals out of the earth. Peoples have gone down before the lessening fertility of the land, and in all probability other peoples will yet go down. The course of empire has been toward the unplundered lands.

The soil crust.

Thinner than a skin of an apple is the covering of the earth that a man tills. The marvelously slight layer that the farmer knows as "the soil," supports all plants and all men, and makes it possible for the globe to sustain a highly developed life. Beyond all calculation and all comprehension are the powers and the mysteries of this soft outer covering of the earth. For all we know, the stupendous mass of materials of which the planet is composed is wholly dead, and only on the surface does any nerve of life quicken it into a living sphere. And yet, from this attenuated layer have come numberless generations of giants of forests and of beasts, perhaps greater in their combined bulk than all the soil from which they have come; and back into this soil they go, until the great life principle catches up their disorganized units and builds them again into beings as complex as themselves.

The general evolution of this soil is toward greater powers; and yet, so nicely balanced are these powers that within his lifetime a man may ruin any part of it that society allows him to hold; and in despair he throws it back to nature to reinvigorate and to heal. We are accustomed to think of the power of man in gaining dominion over the forces of nature,— he bends to his use the expansive powers of steam, the energy of the electric current, and he ranges through space in the light that he concentrates in his telescope; but while he is doing all this, he sets at naught the powers in the soil beneath his feet, wastes them, and deprives himself of vast sources of energy. Man will never gain dominion until he learns from nature how to maintain the augmenting powers of the disintegrating crust of the earth.

We can do little to control or modify the atmosphere or the sunlight; but the epidermis of the earth is ours to do with it much as we will. It is the one great earth-resource over which we have dominion. The soil may be made better as well as worse, more as well as less; and to save the producing powers of it is far and away the most important consideration in the conservation of natural resources.

Unfortunately, it is impossible to devise a system of farm accounting that shall accurately represent the loss in producing power of the land (or depreciation in actual capital stock). The rising sentiment on the fertility question is just now reflected in the proposal to ask Congress and the states to make it a misdemeanor for a man to rob his land and to lay out for him a farm scheme. This is a chimerical notion; but the people are bound to express themselves unmistakably in some way on this subject.

Even if we should ultimately find that crops do not actually deplete land by the removal of stored plant-food in the way in which we have been taught, it is nevertheless true that poor management ruins its productivity; and whatever the phrase we use in our speaking and writing, we shall still need to hold the land-usurer to account.

No man has a right to plunder the soil.

The man who tills and manages the soil owes a real obligation to his fellow-men for the use that he makes of his land; and his fellowmen owe an equal obligation to him to see that his lot in society is such that he will not be obliged to rob the earth in order to maintain his life. The natural resources of the earth are the heritage and the property of every one and all of us. A man has no moral right to skin the earth, unless he is forced to do it in sheer self-defense and to enable him to live in some epoch of an unequally developed society; and if there are or have been such social epochs, then is society itself directly responsible for the waste of the common heritage. We have given every freeholder the privilege to destroy his farm.

The man who plunders the soil is in very truth a robber, for he takes that which is not his own, and he withholds bread from the mouths of generations yet to be born. No man really owns his acres: society allows him the use of them for his lifetime, but the fee comes back to society in the end. What, then, will society do with those persons who rob society? The pillaging land-worker must be brought to account and be controlled, even as we control other offenders.

I have no socialistic program to propose. The man who is to till the land must be educated: there is more need, on the side of the public welfare, to educate this man than any other man whatsoever (page 36). When he knows, and his obligations to society are quickened, he will be ready to become a real conservator; and he will act energetically as soon as the economic pressure for land-supplies begins to be acute. When society has done all it can to make every farmer a voluntary conservator of the fatness of the earth, it will probably be obliged to resort to other means to control the wholly incompetent and the recalcitrant; at least, it will compel the soil-robber to remove to other occupation, if economic stress does not itself compel it. We shall reach the time when we shall not allow a man to till the earth unless he is able to leave it at least as fertile as he found it.

I do not think that our natural soil resources have yet been greatly or permanently depleted, speaking broadly; and such depletion as has occurred has been the necessary result of the conquest of a continent. But a new situation will confront us, now that we see the end of our raw conquest; and the old methods cannot hold for the future. The conquest has produced great and strong folk, and we have been conserving men while we have been free with our resources. In the future, we shall produce strong folk by the process of thoroughness and care.

Ownership vs. *conservation.*

This discussion leads me to make an application to the conservation movement in general. We are so accustomed to think of privileged interests and of corporation control of resources that we are likely to confuse conservation and company ownership. The essence of conservation is to utilize our resources with no waste, and with an honest care for the children of all the generations. But we state the problem to be the reservation of our resources for all the people, and often assume that if all the resources were in private ownership the problem would thereby be solved; but, in fact, the conservation question is one thing and the ownership of property quite another. A corporation may be the best as well as the worst conservator of resources; and likewise, private or individual ownership may be the very worst as well as the best conservator. The individual owner, represented by the "independent farmer," may be the prince of monopolists, even though his operations compass a very small scale. The very fact that he is independent and that he is intrenched behind the most formidable of all barriers — private property rights — insure his monopoly.

In the interest of pure conservation, it is just as necessary to control the single men as the organized men. In the end, conservation must deal with the separate or the individual man; that is, with a person. It matters not whether this person is a part of a trust, or lives alone a hundred miles beyond the frontier, or is the owner of a prosperous farm, — if he wastes the heritage of the race, he is an offender.

We are properly devising ways whereby the corporation holds its property or privileges in trust, returning to government (or to society) a fair rental; that is, we are making it responsible to the people. What shall we do with the unattached man, to make him also responsible? Shall we hold the corporate plunderer to strict account, and let the single separate plunderer go scot-free?

The philosophy of saving.

The conservation of natural resources, therefore, resolves itself into the philosophy of saving, while at the same time making the most and best progress in our own day. We have not developed much consciousness of saving when we deal with things that come free to our hands, as the sunshine, the rain, the forests, the mines, the streams, the earth; and the American has found himself so much in the

Country Life and Conservation 193

midst of plenty that saving has seemed to him to be parsimony, or at least beneath his attention. As a question of morals, however, conscientious saving represents a very high development. No man has a right to waste, both because the materials in the last analysis are not his own, and because some one else may need what he wastes. A high sense of saving ought to come out of the conservation movement. This will make directly for character-efficiency, since it will develop both responsibility and regard for others.

The irrigation and dry-farming developments have a significance far beyond their value in the raising of crops: they are making the people to be conservators of water, and to have a real care for posterity.

Civilization, thus far, is built on the process of waste. Materials are brought from forest, and sea, and mine, certain small parts are used, and the remainder is destroyed (page 20); more labor is wasted than is usefully productive; but what is far worse, the substance of the land is taken in unimaginable measure, and dumped

o

wholesale into endless sewer and drainage systems. It would seem as if the human race were bent on finding a process by which it can most quickly ravish the earth and make it incapable of maintaining its teeming millions. We are rapidly threading the country with vast conduits by which the fertility of the land can flow away unhindered into the unreachable reservoirs of the seas.

The conservation of food.

The fundamental problem for the human race is to feed itself. It has been a relatively easy matter to provide food and clothing thus far, because the earth yet has a small population, and because there have always been new lands to be brought into requisition. We shall eliminate the plagues and the devastations of war, and the population of the earth will tremendously increase in the centuries to come. When the new lands have all been opened to cultivation, and when thousands of millions of human beings occupy the earth, the demand for food will constitute a problem which we scarcely apprehend to-day. We shall then be obliged to develop self-sustaining methods of maintaining the producing-power of land.

We think we have developed intensive and perfected systems of agriculture; but as a matter of fact, and speaking broadly, a permanent organized agriculture is yet unknown. In certain regions, as in Great Britain, the producing-power of the land has been increased over a long series of years, but this has been accomplished to a great extent by the transportation of fertilizing materials from the ends of the earth. The fertility of England has been drawn largely from the prairies and plains of America, from which it has secured its food supplies, from the guano deposits in islands of the seas, from the bones of men in Egypt and the battlefields of Europe.

We begin to understand how it is possible to maintain the producing-power of the surface of the earth, and there are certain regions in which our knowledge has been put effectively into operation, but we have developed no conscious plan or system in a large way for securing this result. It is the ultimate problem of the race to devise a permanent system of agriculture. It is the greatest question that can confront mankind; and the question is yet all unsolved.

The best husbandry is not in the new regions.

The best agriculture, considered in reference to the permanency of its results, develops in old regions, where the skinning process has passed, where the hide has been sold, and where people come back to utilize what is left. The skinning process is proceeding at this minute in the bountiful new lands of the United States; and in parts of the older states, and even also in parts of the newer ones, not only the skin but the tallow has been sold.

We are always seeking growing-room, and we have found it. But now the Western civilization has met the Eastern, and the world is circumferenced. We shall develop the tropics and push far toward the poles; but we have now fairly discovered the island that we call the earth, and we must begin to make the most of it.

Another philosophy of agriculture.

Practically all our agriculture has been developed on a rainfall basis. There is ancient irrigation experience, to be sure, but the great agriculture of the world has been growing away from these regions. Agriculture is still moving on, seeking new regions; and it is rapidly invading regions of small rainfall.

About six-tenths of the land surface of the globe must be farmed, if farmed at all, under some system of water-saving. Of this, about one-tenth is redeemable by irrigation, and the remainder by some system of utilization of deficient rainfall, or by what is inappropriately known as dry-farming. The complementary practices of irrigation and dry-farming will develop a wholly new scheme of agriculture and a new philosophy of country life (page 44).

Even in heavy rainfall countries there is often such waste of water from run-off that the lands suffer severely from droughts. No doubt the hilly lands of our best farming regions are greatly reduced in their crop-producing power because people do not prepare against drought as consciously as they provide against frost (page 52). It is often said that we shall water Eastern lands by irrigation, and I think that we shall; but our first obligation is to save the rainfall water by some system of farm-management or dry-farming.

Agriculture rests on the saving of water.

The obligation of the farmer.

The farmer is rapidly beginning to realize his obligation to society. It is usual to say that the farmer feeds the world, but the larger fact is that he saves the world.

The economic system depends on him. Wall Street watches the crops.

As cities increase proportionately in population, the farmer assumes greater relative importance, and he becomes more and more a marked man.

Careful and scientific husbandry is rising in this new country. We have come to a realization of the fact that our resources are not unlimited. The mining of fertilizing materials for transportation to a few spots on the earth will some day cease. We must make the farming sustain itself, at the same time that it provides the supplies for mankind.

We all recognize the necessity of the other great occupations to a well-developed civilization; but in the nature of the case, the farmer is the final support. On him depends the existence of the race. No method of chemical synthesis can provide us with the materials of food and clothing and shelter, and with all the good luxuries that spring from the bosom of the earth.

I know of no better conservators than our best farmers. They feel their responsibility. Quite the ideal of conservation is illustrated by a farmer of my acquaintance who saves every product of his land and has developed a system of self-enriching live-stock husbandry, who has harnessed his small stream to light his premises and do much of his work, who turns his drainage waters into productive uses, and who is now troubled that he cannot make some use of the winds that are going to waste on his farm.

The obligation of the conservation movement.

What I have meant to impress is the fact that the farmer is the ultimate conservator of the resources of the earth. He is near the cradle of supplies, near the sources of streams, next the margin of the forests, on the hills and in the valleys and on the plains just where the resources lie. He is in contact with the original and raw materials. Any plan of conservation that overlooks this fact cannot meet the situation. The conservation movement must help the farmer to keep and save the race.

PERSONAL SUGGESTIONS

In the preceding pages I have tried to develop the reader's point of view. To do this, I have gone over very briefly some of the questions that are now actively under discussion. There are other matters, of a more personal nature, that need to be discussed, or at least mentioned, in connection with even a sketchy consideration of the country-life movement; and some of these I now place together in a closing budget.

The open country must solve its own problems.

It may first be said that the reconstruction of the open country must depend in the main on the efforts of the country people themselves. We are glad of all interchange of populations; the influx of country blood has thus far been invaluable to cities; the outgo of city people has set new aspirations into the country, and it is still necessary to call on the cities for labor in times of pressure: but stated in its large terms, the open country will rise no higher than the aspirations of the people who live there, and the problems must be solved in such way that they will meet the conditions as they exist on the spot.

Profitable farming is not a sufficient object in life.

It may then be said that it is the first duty of every man to earn a decent living for himself and those dependent on him; and a countryman cannot expect to have much influence on his time and community until he makes his farm pay in dollars and cents.

But the final object in life is not to make money, but to use money in developing a higher type of endeavor and a better neighborhood. The richest farming regions do not necessarily have the best society or even the best living conditions. Social usefulness must become a fact in country districts. The habit of life in the usual farm family is to take everything to itself and to keep it. Standards of service must take the place of standards of property.

New country professions.

The country-life movement does not imply that all young persons who hereafter shall remain in the country are to be actual farmers. The practice of customary professions and occupations will take on more importance in country districts. The country physician, veterinary, pastor, lawyer, and teacher are to extend greatly in influence and opportunity.

But aside from all this, entirely new occupations and professions are to arise, even the names of which are not yet known to us. Some of them are already under way. There will be established out in the open country plant doctors, plant-breeders, soil experts, health experts, pruning and spraying experts, forest experts, farm machinery experts, drainage and irrigation experts, recreation experts, market experts, and many others. There will be housekeeping experts or supervisors. There will be need for overseers of affiliated organizations and stock companies. These will all be needed for the purpose of giving special advice and direction (page 78). We shall be making new applications of rural law, of business methods for agricultural regions, new types of organization. The people will find that it will pay to support such professions or agents as these.

Country life will become more complex as rapidly as it becomes more efficient.

The personal resources.

The attitude toward one's world has much to do both with his effectiveness and with his satisfaction in living; and this is specially true with the farmer, because he is so much alone and has so few conventional sources of entertainment. It may be important to provide new entertainment for the farmer; but it is much more important to develop his personal resources.

The simple life, as Pastor Wagner so well explains, is a state of mind. It is a simplification of desire, a certain directness of effort and of purpose that brings us quickly to a result, and such an attitude that we derive our satisfactions from the humble and the near-at-hand. The countryman is the man who has the personal touch with his environment.

With the increasing differentiation of country life, it is of the first importance that the country people do not lose their simplicity of desires.

The meaning of the environment.

It is too little appreciated that every natural object makes a twofold appeal to the human mind: its appeal in the terms of its physical or material uses, and its appeal to our sense of beauty and of personal satisfaction. As the people progresses in evolution, the public mind becomes constantly more sensitive to the conditions in which we live, and the appeal to the spiritual satisfaction of life constantly becomes stronger. Not only shall the physical needs of life be met, but the earth will constantly be made a more satisfactory place in which to live.

We must not only save our forests in order

that they may yield timber and conserve our water supplies, but also that they may adorn and dominate the landscape and contribute to the meaning of scenery. It is important that our coal supplies be conserved not only for their use in manufacture and the arts, but also that smoke does not vitiate the atmosphere and render it unhealthful, and discolor the objects in the landscape. It is of the greatest importance that water supplies be conserved by storage reservoirs and other means, but this conservation should be accomplished in such a way as not to menace health or offend the eye or destroy the beauty of contiguous landscape. The impounding of waters without regard to preserving natural water-falls, streams, and other scenery, is a mark of a commercial and selfish age, and is a procedure that cannot be tolerated in a highly developed society. It is important that regulations be enacted regarding the operation of steam roads through woody districts, not only that the timber may be saved, but also that the natural beauty of the landscape may be protected from fire and other forms of destruction. The fertility of the soil must be saved, not only that products may be raised with which to feed and clothe the people, but also that the beauty of thrifty and productive farms may be saved to the landscape. The property-right in natural scenery is a tenure of the people, and the best conservation of natural resources is impossible until this fact is recognized.

On this point the Commission on Country Life makes the following statement: " In estimating our natural resources we must not forget the value of scenery. This is a distinct asset, and it will be more recognized as time goes on. It will be impossible to develop a satisfactory country life without conserving all the beauty of landscape and developing the people to the point of appreciating it. In parts of the East, a regular system of parking the open country of the entire state is already begun, constructing the roads, preserving the natural features, and developing the latent beauty in such a way that the whole country becomes part of one continuing landscape treatment. This in no way interferes with the agricultural utilization of the land, but rather increases it. The scenery is, in fact, capitalized, so that it adds to the property values and contributes to local patriotism and to the thrift of the commonwealth."

Historic monuments.

The general tendency of our time is to dump everything into the cities, particularly into the large cities. It is there that we assemble our treasures of art, our libraries, our dramatic skill, our specimens of statuary and architecture; and it is there that the aspiring men also assemble to work out their destinies. And yet there have been events in the open country. Great men have lived there. Things have come to pass. We should be interested to record these events of the rural country, as well as the events that are associated with the congested city. Persons of quickened intelligence will not live contentedly in the outer country if it provides nothing more than subsistence. Every new memorial in the farming country is one additional reason for people to live there.

The open country as well as the city has a history; but one would not discover the fact from monuments that he may see.

It may not be possible now to erect elaborate monuments far in the country to commemorate historical events, but records may be made, and it is at least possible to roll up a pile of stones.

Improvement societies.

Of late years there has sprung up a line of societies in villages and small cities whose province it is to create public sentiment for the betterment of the place in general good looks, and which, for lack of a better name, are generally collectively known as "village improvement societies." These organizations have had much effect in making the villages attractive. Their influence extends far and wide, but the organization itself in any case ought to take in all the surrounding territory, with the purpose to secure a coöperative action between town and country (page 122). The entire region, not city or town alone, should be organized.

In many rural communities, there could well be an open-country improvement society; or an organization might be formed, from the church or otherwise, to care for a particular interest, as the school ground or the cemetery. The average country cemetery particularly needs attention.

The care of all the public or semi-public property of a township or a neighborhood is somebody's responsibility, and this responsibility should be recognized in organization. The pride of the community could be greatly stimulated if a group of people should associate to look after roadsides, lake shores and river banks, waste places, deserted and dilapidated buildings, weeds, raw spots, paths, dangerous places, mosquito ponds, breeding places of insects, stray dogs, horse sheds, trees, birds, wild flowers, guide-posts, telephone depredations, cemeteries, church grounds, school grounds, almshouse grounds, picnic grounds, historic places, patriotic events, bits of good scenery, and to give advice on lawns, back-yards and barn-yards, advertising signs.

Entertainment.

All persons seem to be agreed that more entertainment and recreation should be provided for country residents; but it does not follow that vaudeville, and the usual line of moving pictures, and the traveling concert would add anything really worth while, although these are often recommended by town folk. The Board Walk kind of pageant may very well be left at the sea-shore.

But we certainly need entertainment that will help country people over the hard and dry places, and raise their lives out of monotony. The guiding principles are two: an entertainment that shall express the best that there is in country life; one that shall set the people themselves at work to produce it, rather than to bring it in bodily from the outside.

I would not eliminate good things merely because they come from the outside, and no one would deny the countryman the touch with any of the masterpieces; but I am speaking now of a form of effort that shall quicken an entire country district and leave a permanent impression on it. I would rather leave the situation as it is than to introduce the meaningless performances of the city thoroughfare and the resorts.

The movement to provide new and better sports, games, and general recreation is now well under way, and I do not need to explain it here; but two things ought to begin to receive attention: music and drama.

The *music spirit* seems to be dying out in the country. I hear very little joyous song there, even though the people may be joyous. The habit of self-expression in song and music needs much to be encouraged in home and school and grange and church. I think the lack is in part due to the over-mastering influence of professional town music, and in part to the absence of study of simple country forms. Simplicity is not now the fashion in music. The single player with a simple theme and the single singer with a melodious and un-

Personal Suggestions 213

trilled strain are not much heard at gatherings now. Some of the best singing I hear is now and then out among the folk,—a simple direct song as plain and sweet as a bird's note. I hope we shall not lose it.

A *drama* of some kind is very much needed for country districts. It should be a new form, something in the way of representing the end of the planting, the harvest, the seasons, the leading crops, the dairy, the woods, the history and traditions of the neighborhood or the region. Many of the pieces should be acted out of doors, and they should be produced chiefly by local talent. Such simple plays for the most part need yet to be written, but the themes are numerous. Why not have a festival or a generous spectacle of Indian corn, and then fill the whole occasion full of the feeling of the corn? As pure entertainment, this would be worth any number of customary theatricals, and as a means of bringing out the talent of the community it would have very positive social value. The traveling play usually leaves nothing behind it.

The themes for short, simple, and strong dramatic presentation are almost numberless,— such episodes and events, for example, as the plowing, the reaping, the husking, the horse-shoeing, the hay-stacking, the wood-chopping, the threshing, the sugaring, the raising of the barn, the digging of the well, the herding of the cattle, the felling of the tree, the building of the church, the making of the wagon, the bridging of the creek, the constructing of the boat, the selling of the farm, the Indians, the settlers, the burst of spring, the dead of winter, the season of bloom, the heyday of summer.

We do not sufficiently appreciate how widespread and native is the desire to dramatize. The ritual of fraternal orders is an illustration. We see it in the charades of evening parties. The old school "exhibition" made a wonderful appeal. Every community likes to see its own people "take parts." At nearly every important grange meeting, and at other country meetings, some one must "recite," and the recitation usually has characters, situations, and "take-offs." It is too bad that we do not have better literature to put in the hands of these reciters; in the meantime, I hope that the custom will not die out.

One who has seen the consummate Passion Play at Oberammergau must have had the thought impressed on him that there is much latent talent among the country folk, and also that it is much worth the while of a community to develop this talent. Aside from its transcendent theme, this stupendous play appeals to the world because of its simplicity and directness and because of its reality, for these are the very kind of folk that might have taken part in the mighty drama had the Great Master lived in Oberammergau.

The nativeness of the play impresses one. The very absence of so much that we associate with the ordinary drama gives the play an appeal,— the absence of the studied stride and strut, of the exaggerated make-ups, and of the over-doing of the parts. The play is grounded in the lives of the people in the community.

We cannot expect another place to become

an Oberammergau, but it is possible for something good to come out of any spot. This thought is vividly expressed by W. T. Stead in his account of the Passion Play:

"As I write, it is now two days after the Passion Play. The crowd has departed, the village is once more quiet and still. The swallows are twittering in the eaves, the blue and cloudless sky over-arches the amphitheater of hills. All is peace, and the whole dramatic troupe pursue with equanimity the even tenor of their ordinary life. Most of the best players are woodcarvers; the others are peasants or local tradesmen. Their royal robes or their rabbinical costumes laid aside, they go about their ordinary work in the ordinary way as ordinary mortals. But what a revelation it is of the mine of latent capacity, musical, dramatic, intellectual, in the human race, that a single mountain village can furnish, under a capable guidance, and with adequate inspiration, such a host competent to set forth such a play from its tinkers, tailors, plowmen, bakers, and the like! It is not native capacity that is lacking to mankind. It is the guiding brain, the patient love, the careful education, and the stimulus and inspiration of a great idea. But, given these, every village of country yokels from Dorset to Caithness might develop artists as noble and as devoted as those of Oberammergau."

The business of farming.

After all is said and done, the first question still remains,— the opportunity to make a good living on a farm, and the possibility of leading a life that will be personally satisfactory.

There has never been a time when farming as a whole has been so prosperous as now, notwithstanding the fact that there are hardships in many regions. The whole occupation is undergoing a process of readjustment, and it is natural that the readjustment has become more complete and perfect in some places and in some kinds of farming than in others. We have but recently passed through a time in which the farming business, except in special regions or special cases, could not be really profitable and attractive.

To make a good and satisfactory living on the farm is a matter both of temperament and of first-class training. There are great series of city vocations in which any person with fair ability can succeed; but farming is a personal business and each man is his own manager. No one should ever go into farming impersonally.

Many persons are making a comfortable living on farms, a better living in fact than persons of similar ability and expending similar energy are making in town. Other persons are failing.

I am not advising anybody to establish himself in the open country; but I am saying that the time has now come when good talent need not avoid the open country.

This is a good time for the well-trained farm-minded young man or woman to go into agriculture; but one should be sure that he has the qualifications.

There is no need that farming provide only a narrow and deadening life. One may express there all the resources of a good education.

The college man is now beginning to affect the sentiment and the practice in rural communities. Formerly a college man going back to the farm was likely to be the subject of distrust and even ridicule. This attitude is passing very rapidly in the good rural regions.

In his public relations, most of the ambition of the countryman has been to hold office. It is a form of small political entertainment, too often with no thought of any particular service to the community. We have a wholly distorted idea of the "honor" of holding office; there is no honor in an office unless it contributes something worth while to society. We cannot expect strong leadership to develop in the open country until there are better things to look forward to than merely to hold the small political places. Many opportunities for rendering prominent public service will now arise in the farm country; perhaps this book will suggest a few of them. And it ought to be some satisfaction to a young man or woman to know that he or she is part of a world-movement, and to feel that it is no longer necessary to explain or to apologize for being a countryman or a farmer.

We have been living in a get-rich-quick age. Persons have wanted to make fortunes. Our business enterprises are organized with that end in view. Persons are now asking how they may live a satisfactory life, rather than placing the whole emphasis on the financial turnover of a business. There is greater need of more good farmers than of more millionaires.

My reader may wish to know what constitutes a good farmer. I think that the requirements of a good farmer are at least four:

The ability to make a full and comfortable living from the land;

to rear a family carefully and well;

to be of good service to the community;

to leave the farm more productive than it was when he took it.